St Antony's Series

Series Editors
Dan Healey, St. Antony's College, University of Oxford, Oxford, UK
Leigh Payne, St. Antony's College, University of Oxford, Oxford, UK

The St Antony's Series publishes studies of international affairs of contemporary interest to the scholarly community and a general yet informed readership. Contributors share a connection with St Antony's College, a world-renowned centre at the University of Oxford for research and teaching on global and regional issues. The series covers all parts of the world through both single-author monographs and edited volumes, and its titles come from a range of disciplines, including political science, history, and sociology. Over more than forty years, this partnership between St Antony's College and Palgrave Macmillan has produced about 400 publications.

This series is indexed by Scopus.

Samuel León Sáez

Mexico's Fuel Trafficking Phenomenon

Analysing an Emerging Black Market

Second Edition

Samuel León Sáez
University of Oxford
Oxford, UK

ISSN 2633-5964 ISSN 2633-5972 (electronic)
St Antony's Series
ISBN 978-3-031-70502-1 ISBN 978-3-031-70503-8 (eBook)
https://doi.org/10.1007/978-3-031-70503-8

© The Editor(s) (if applicable) and The Author(s), under exclusive license to Springer Nature Switzerland AG 2021, 2025

This work is subject to copyright. All rights are solely and exclusively licensed by the Publisher, whether the whole or part of the material is concerned, specifically the rights of translation, reprinting, reuse of illustrations, recitation, broadcasting, reproduction on microfilms or in any other physical way, and transmission or information storage and retrieval, electronic adaptation, computer software, or by similar or dissimilar methodology now known or hereafter developed.
The use of general descriptive names, registered names, trademarks, service marks, etc. in this publication does not imply, even in the absence of a specific statement, that such names are exempt from the relevant protective laws and regulations and therefore free for general use.
The publisher, the authors and the editors are safe to assume that the advice and information in this book are believed to be true and accurate at the date of publication. Neither the publisher nor the authors or the editors give a warranty, expressed or implied, with respect to the material contained herein or for any errors or omissions that may have been made. The publisher remains neutral with regard to jurisdictional claims in published maps and institutional affiliations.

Cover credit: © Matteo Legrenzi, 2023

This Palgrave Macmillan imprint is published by the registered company Springer Nature Switzerland AG
The registered company address is: Gewerbestrasse 11, 6330 Cham, Switzerland

If disposing of this product, please recycle the paper.

For Carmen and Yara

Acknowledgements

This research went through many stages and along each one different people were of enormous help to make it the piece of work it is today. I would like to thank my interview respondents who shared their expertise and insights with wonderful generosity: Ana Lilia Pérez, Gustavo Mohar, Edmundo Velázquez, Eduardo Guerrero, Alejandro Hope, Rubén Salazar, Sergio Mastretta, Dante San Pedro and Gabriel Stargardter. During the interview process the help of my brother Santiago León was of enormous importance and I am deeply grateful for it. I would also like to thank Helena Varela, Juan Federico Arriola and Rubén Aguilar for giving me their support at the beginning of this journey. At the University of Oxford, I had the wonderful support of brilliant academics like Carlos Pérez Ricart, Leigh Payne, Laurence Whitehead and Carlos Solar who helped to make this investigation what it is today. I want to thank each one of them for their interest, their valuable advice and the time they generously invested in my research. I am also deeply grateful for the help of Adrián Lajous, Carlos Elizondo Mayer-Serra and Pedro Sáez, who helped me to clarify important themes of this research in a more advanced phase. Finally, I would like to thank Yara Tarabulsi and Carmen Sáez for their vital support, for every enriching conversation, for each time they listened and for each wonderful piece of advice they gave along the way.

CONTENTS

Part I   An Ambush

1   Introduction   3
    Introduction   3
    About This Research   7
    A Mexican Case Study, with Far Reaching Implications   11
    Bibliography   12

2   An Introduction to Fuel Trafficking   15
    A Global Introduction to Fuel Trafficking   15
    Researching and Defining Fuel Trafficking   15
    Fuel Trafficking: A Global Criminal Market   16
    Of Fuel Trafficking, Taxes and Subsidies   17
    Criminal Networks of All Sizes Participate in Fuel Trafficking   18
    Grey Actors: An Unescapable Element in Fuel Trafficking   20
    Fuel Trafficking: Camouflaging Illegality   22
    Theorizing Mexican Fuel Trafficking   23
    Diversification and Criminal Fragmentation   23
    Black Markets: A Theoretical Overview   27
    The Fuel Black Market and the Co-Option of PEMEX   31
    The Maltese Connection   33
    Introduction   34
    The Libyan Security Context   34

The Libyan Energy Context	34
The Libyan Fuel Black Market	36
The Libyan-Maltese Connection	37
The Network: Actors and Functions	38
The Continental Actors	45
The Lessons of the Ben Khalifa/Debono Network	46
Bibliography	50

Part II   La Sierra Norte de Puebla

3   The Mexican Context 59

Mexico's Energy and Security: A Tale of Two Crises	59
The Energy Crisis	59
The Security Crisis	62
The Peña Nieto Years	66
The Origins of Mexican Fuel Trafficking	68
The Río Grande Valley Revelation	71
The Burgos Basin Operation	71
Burgos: The Mexican Side	75
Explaining the MFBM	79
The Relevance of the MFBM	79
MFBM Operations: Installations and Pipelines	83
Theft in Refineries and Storage and Distribution Terminals	85
Pipeline Theft	86
Maritime Fuel Trafficking	90
The Incentives of the MFBM	94
The Actors in the MFBM	96
Bibliography	100

Part III   A Journey from Monterrey to Mexico City

4   Understanding the Growth of the Mexican Fuel Black Market 109

Factors Behind Fuel Trafficking Growth in Mexico	109
Fragmentation and Diversification of Criminal Networks	110
Criminal Dynamics: Fragmentations and Alliance Formation	111
Diversification Towards New Criminal Markets	112
The Zetas: Pioneering Black-market Diversification	113
The Zetas as Fuel Traffickers	115

Mexico As a Criminal Hub 117
Mexico's Criminal Landscape 118
Specialized Fuel Trafficking Networks Affiliated to Larger
  Criminal Groups 120
Co-option of Grey Actors 122
PEMEX: A Breeding Ground for Co-option 122
Oil Workers: Crucial Grey Actors 123
The Union and the Plunder 124
White Collars and Bureaucrats 125
The Distribution Network 126
Grey Actors and the Co-opted Institutional
  Reconfiguration of PEMEX 127
Fuel Prices Increases in Mexico 128
Bibliography 132

Part IV Guanajuato and the Dispute for Salamanca

5 The Local Dynamics of Fuel Trafficking in Puebla and Guanajuato 141
Introduction 141
Vignette Four: Guanajuato and the Dispute for Salamanca 142
  Energy Installations and Criminal Violence in Guanajuato 144
  The Criminal Networks Behind the MFBM in Guanajuato 145
  Guanajuato: Co-opting Local Authorities 149
The Puebla Case: Of Development and Criminal Geographies 150
  Community Participation in the Poblano MFBM 151
  The Development Hubs 154
  Criminal Networks Behind the MFBM in Puebla 156
  The Grey Actors in Puebla's MFBM 158
The MFBM in Guanajuato and Puebla 160
Bibliography 162

6 The Mexican Fuel Black Market After a One Year Crackdown 169
Introduction 169
The AMLO Energy Context 169
The Obrador Security Context 171
  Of Hugs, State Lethal Force and Homicidal Violence 174
  Criminal Homicidal Violence Under AMLO 175
  Triumphalism and the Crackdown(s) on the MFBM 176

    Of Local Dynamics, Inconsistencies and the Case Against Anticipated Triumphalism   179
    SITRAC and the Co-opted Institutional Reconfiguration of PEMEX   185
    Closing Pipelines: A Costly Solution   189
    The Resiliency of Mexican Criminal Networks   193
    The Diversification of the Fuel Black-Market   194
    The Motivations to Diversify Towards LPG   196
    The MFBM After One Year of the Crackdown   201
    Bibliography   204

7 Conclusion   213
    Bibliography   219

Index   221

Abbreviations

AMLO	Andrés Manuel López Obrador
bpd	Barrels Per Day
CItR	Co-opted Institutional Reconfiguration
CJNG	*Cártel Jalisco Nueva Generación*
CStR	Co-opted State Reconfiguration
EPN	Enrique Peña Nieto
GC	Gulf Cartel
MFBM	Mexican Fuel Black Market
MXN	Mexican Peso
PAN	National Action Party
PEMEX	*Petróleos Mexicanos*
PRD	Democratic Revolution Party
PRI	Institutional Revolutionary Party
SOE	State Owned Enterprise (*Petróleos Mexicanos*)
USD	United States Dollar

List of Figures

Fig. 3.1	National homicides vs totals excluding Chihuahua, Coahuila and Durango 2010–2014 (Hope 2017)	66
Fig. 3.2	Extraction points PEMEX pipelines 1996–2001 (Córdoba 2003; Cervantes 2012)	69
Fig. 3.3	Extraction points PEMEX pipeline network (PEMEX: "Reporte de tomas clandestinas"; EnergeA and Grupo Atalaya 2017, 344; Transparency Petition Response 1857000059819)	80
Fig. 3.4	PEMEX non-operating losses due to illicit activities ("Annual Report Pursuant…" 2018a, 319)	82
Fig. 3.5	Stolen barrels in PEMEX: installations vs pipelines (2013–2018) (Romero 2019a)	84
Fig. 4.1	Evolution of Hydrocarbon Prices in Mexico 2009–2017 (per litre) NB *(Northern Border)* RC *(Rest of the Country) (Secretaría* de Energía 2018, 44)	129
Fig. 4.2	From Subsidies to Taxation: Evolution of the IEPS during the Calderón and the Peña Nieto Administrations *(Elaborated by author with information of Moreno 2017)*	130
Fig. 4.3	Stolen Barrels vs Extraction Points (2010–2018) *(Sánchez 2012;* EnergeA and Grupo Atalaya 2017, 344*; Romero 2019a;* PEMEX: *"Reporte de tomas clandestinas")* *Official data for stolen barrels reported in 2016 did not include pipeline losses	131

Fig. 5.1	Illegal Extraction Points in Guanajuato (2012–2018) (Transparency Petition Response 1857200119616; *Observatorio Nacional Ciudadano* 2018a, 7)	143
Fig. 5.2	Illegal extraction points in Puebla (2011–2018) (Gobierno Fácil 2019; PEMEX: "Reporte de tomas clandestinas")	151
Fig. 6.1	Areas of security budget spending 2018 vs 2019 (%) ("La guerra que sigue" 2020)	172
Fig. 6.2	Extraction points PEMEX Pipeline Network (PEMEX: "Reporte de tomas clandestinas"; EnergeA and Grupo Atalaya 2017, 344; Transparency Petition Responses 1857000059819 and 1857200094319)	180
Fig. 6.3	Criminal homicides and illegal pipeline extraction points in Guanajuato 2009–2019 (*Source* SESNSP; López et al. 2020; Transparency Petition Responses 1857200007220, 1857000059819 and 1857200094319)	183
Fig. 6.4	Fuels missing on PEMEX's Custody Transfer metering system (SITRAC) 2019 (Thousands of Barrels) (*Source* Transparency Petition Response 1857000006220)	186
Fig. 6.5	Fuel types missing on SITRAC system 2019 (Barrels) (*Source* Transparency Petition Response 1857000006220)	187
Illustration 1.1	Research theoretical framework and analytical tools	10
Illustration 2.1	Certificates of origin used by the Ben Khalifa/Debono Network (Guardia Di Finanza Di Catania 2017)	44
Illustration 2.2	Actors and functions of the Ben Khalifa/Debono Network (Elaborated by author)	47
Illustration 3.1	Fuel trafficking scheme in the Burgos Basin. *Source* Elaborated by author with information from Garay-Salamanca and Salcedo-Albarán 2016, 96–110	74
Illustration 3.2	Fuel-theft in refineries and distribution terminals (Elaborated by author with information from Pérez 2018a)	86
Illustration 3.3	PEMEX pipeline management areas (Elaborated by author with information of Flores 2019)	89

Illustration 3.4	MFBM criminal network typology (Elaborated by author with information from EnergeA and Grupo Atalaya 2017, 19–21)	100
Illustration 4.1	The Mexican Fuels Black Market (MFBM) (Elaborated by the author)	129
Illustration 6.1	Liquified petroleum gas value chain and points of criminal network intervention (Elaborated by the author with information from Madrid Ayala et al. 2018, 30)	200
Map 3.1	Energy infrastructure in the Burgos Basin (Llano and Flores 2017)	73
Map 3.2	Polyduct systems and regions with high concentration of illegal extraction points 2009–2016 (EnergeA and Grupo Atalaya 2017, 38)	88
Map 3.3	PEMEX strategic energy installations (EnergeA and Grupo Atalaya 2017, 9)	91
Map 5.1	Energy Installations in Guanajuato (Llano and Flores 2017)	144
Map 5.2	Criminal conflict and criminal incidence in Guanajuato (Elaborated by author with information of Saucedo 2019)	148
Map 5.3	Municipal fuel trafficking hotspots and energy installations in Puebla	153
Map 5.4	MFBM consuming hubs in Puebla (Elaborated by author with information of Mastretta 2018)	156
Map 6.1	Comparison of Mexico's gas pipeline network vs Polyduct network (Llano and Flores 2017)	197

List of Tables

Table 2.1	Illegal/Black market types	29
Table 3.1	Stolen PEMEX barrels due to fuel-theft 2013–2018	85
Table 3.2	Fuel-theft in PEMEX pipelines by fuel type 2011–2015	87
Table 3.3	Ranking of polyduct systems by volumetric losses and illegal extraction points 2011–2015	87
Table 4.1	Municipalities with more illegal pipeline extraction points 2009–2016	114
Table 4.2	Fuel trafficking specialized subnetworks	121
Table 6.1	Illegal extraction points in pipelines 2018–2019	179
Table 6.2	Illegal extraction points in pipelines 2018–2019	181
Table 6.3	Selected cases of missing fuel SITRAC system 2019	188
Table 6.4	Selected cases of missing fuel SITRAC system 2019	190
Table 6.5	Illegal extraction points in LPG pipelines 2018–2019	195
Table 6.6	Illegal extraction points in PEMEX LPG pipelines 2018–2019	198
Table 6.7	Legal complaints presented before the Attorney General of the Republic (FGR) for LPG tanker truck theft	199

PART I

An Ambush

The hour is 8:23 pm. Three men are driving down a straight road in the middle of the night on motorbikes. One of them stops in a corner. He is spying. A military convoy appears behind him. Suddenly, two vans accelerate violently enclosing one of the military vehicles, a pick-up truck with troops in its cargo area. Tensions are running high. Two military trucks reach the point of the enclosure. Less than two minutes have passed. The soldiers leave their vehicle and grab one of the men inside the enclosing vans, a civilian wearing a bullet-proof vest. They put him on the ground. Within seconds the soldiers start falling back: they are being shot from the rear-guard.

As they retreat, the bullet-proof vested man shoots a soldier in the back while lying on the floor. The soldier tumbles to the ground, motionless. The soldiers find cover behind one of the ambushers' vans. The scene descends into chaos. Suddenly, a van approaches the soldiers from the attackers' side, who without hesitation respond firing back. Advancing vehicles can be seen in the rear where the van came from; they falter and reverse.

By 8:29 there is a ceasefire. The soldiers take their fallen colleague into cover. The soldiers approach the van that advanced onto them, shattered by the gunshots it received moments before. They force four men out of the vehicle. A fifth one is dragged out, alive but motionless. The military members drop him on the ground and proceed to move the enclosing vans. At 8:44:26, a soldier approaches the motionless attacker on the ground and shoots him point-blank in the back of the skull. By 10:21 pm, the soldiers have secured the area. Once they notice the cameras spying on them, they take them down ("¿Qué Pasó el 3 de mayo..." 2017). This crossfire took place on May 3, 2017, in Mexico, a country where clashes between the military and dangerous criminal groups have become tragically all too common. The site of this incident was a town in the state of Puebla where criminal activity had previously been infrequent. That night, cameras recorded a fraction of a broader confrontation that killed four soldiers and six criminals. Next morning, the videos were uploaded to the internet and Mexico's media exploded in a frenzy: the men involved in the attack were not drug cartel members but fuel traffickers. Without warning, the problem caught public attention.

Reference

"¿Qué Pasó el 3 de mayo en Palmarito, Puebla?" Noticieros Televisa. May 11, 2017. https://noticieros.televisa.com/ultimas-noticias/que-paso-3-mayo-palmarito-puebla/ . Date Accessed July 13, 2021.

CHAPTER 1

Introduction

INTRODUCTION

This investigation focuses on explaining one of the most recent and relevant black markets of Mexico: fuel trafficking. This illegal enterprise has shown a global presence from North to South America, from the Middle East and North Africa to Europe, and from Southeast Asia to Sub-Saharan Africa (Ralby 2017). The theft and illegal commercialization of fuels has become a lucrative market for various criminal agents involved in other illicit extractive industries (rare minerals, timber, gold).

This trend is part of a diversification phenomenon changing the scope of criminal activity worldwide. Criminal networks continuously explore opportunities to diversify into new black markets, assessing potential returns and risks of engaging in different illegal enterprises. Diversification helps these groups *"reduce risks to their commercial portfolios by diversifying into profitable activities with low probability of being detected"* (OECD 2016, 20–21). Expansion into new criminal markets creates sources of revenue (critical in contexts where state repression and criminal competition are rampant) and enables illicit networks to capitalize on their transportation and money-laundering structures. Fuel trafficking allows access to vast revenues while offering important incentives like the exploitation of formal energy supply chains increasing the possibility of avoiding detection.

Across the world, many fuel trafficking cases have stood out. This research focuses on the Mexican case. By analysing Mexican fuel trafficking in detail, this investigation contributes to wider-reaching research on public security and criminality. Black markets in subnational enclaves within Mexico, Central America, Colombia, Venezuela and Brazil have become prized assets for the region's illicit underworld and are the reason behind increasing violent criminal competition (Yashar 2018). These illicit subnational markets share the following traits: (1) they generate high profits, (2) require territorial control and (3) shift geographically (ibid., 68). This type of illicit economies can expand in areas with weak state capacity, widespread corruption, and possibilities of co-option. As illicit economies have expanded, so has the number of criminal actors involved in pursuing territorial control for extractive activities. This research contributes by explaining how an illegal market with strong presence at the state and local level expanded to the point of becoming one of the main illicit enterprises in Mexico. This study thus has implications not only at the national level but also the regional one, shedding light on how these black markets consolidate and thrive.

Fuel trafficking has become one of the main Mexican criminal activities in recent years, reconfiguring the geography of criminality (Hope 2017; Guerrero 2018) and representing a strategic threat to national company *Petróleos Mexicanos* (PEMEX) (*"Informe Anual 2017"* 2018, 74). This research develops the origins of Mexican fuel trafficking and explains its exponential growth between 2011 and 2018. It will also present how this criminal enterprise has changed during the first year of the Obrador (AMLO) presidency (2018–2024).

Fuel trafficking has mostly been researched by journalists (Pérez 2011, 2017; Mastretta 2017). Academic research explaining its origins, workings and growth is yet to be done. There is scarce analysis of the problem, and interpretative frameworks to understand it as a criminal phenomenon are lacking (Esparza and de Paz Mancera 2017). The main questions this research thus answers are: What are the origins of Mexican fuel trafficking? How did it become a prominent criminal market in Mexico and what factors explain its growth between 2011 and 2018?

This research's objectives were determined by data gathered through semi-structured interviews conducted in Mexico in 2018. My respondents were consultants, former intelligence officials, security analysts, journalists, and a former PEMEX official. Their insights direct this research and constitute its primary information sources. Interviews were determined by assessing the literature on the subject which allowed me to identify individuals knowledgeable about Mexican fuel trafficking (Van Audenhove 2017).

Other relevant sources were official documents, reports and journalistic investigations. Many of the works used in this research were elaborated by journalists and specialists I interviewed during fieldwork, building upon their written arguments. This helped contextualize certain characteristics of the fuel trafficking market such as scale, economic and political context, actors involved and history. Official documents obtained through transparency petitions were another source of valuable information. This research builds on different theoretical frameworks on criminality to explain the consolidation of Mexican fuel trafficking, and along the way contributes to a broader discussion on criminal networks, public security and contemporary criminality in Mexico, the Americas and beyond.

The growth of fuel trafficking in Mexico led to the formation of a sophisticated Mexican Fuel Black Market (MFBM). The works of economic anthropologists were relevant to this research's understanding of black markets given their emphasis on the social organization of markets and the actors shaping them (Beckert and Wehinger 2013; Beckert et al. 2015; Beckert and Dewey 2017; Hübschle 2017). This literature helped interpret the MFBM's characteristics, establish typologies and explain local-level phenomena. Other important theoretical frameworks explore the trend of criminal fragmentation and enterprise diversification that consolidated in the Mexican criminal underworld (Garzón 2014; Atuesta and Pérez-Dávila 2018; Durán-Martínez 2018; Correa-Cabrera 2017; Yashar 2018).

Other researchers contribute to the analysis of the expansion of Mexican fuel trafficking by developing theory regarding the co-option of state and non-state actors, as well as institutions. Co-option dynamics are crucial for criminal markets, including fuel trafficking, to thrive (Garay-Salamanca and Salcedo-Albarán 2016; Garay-Salamanca et al. 2018; Lessing 2015). Research on criminal networks contributed to my work by helping define the actors who enabled the growth of the MFBM (Williams 2001; Kenney 2007; Ayling 2009; Bright and Delaney 2013).

When designing this research, my impression was that fuel trafficking in Mexico was growing and impacting the nation's criminal landscape. Yet this perception lacked support as official figures on the issue were indirect (data on the growth of illegal extraction points in pipelines was available, while volumetric losses in litres of stolen fuels were not) and presented multiple limitations. With this information I began my fieldwork.

My fieldwork interview respondents were selected in order to examine the issue as closely as possible without endangering my security as a researcher. Reaching fuel traffickers directly was not viable given the violent nature of this black market (Hobbs and Antonopoulos 2014, 14). Despite these risks, many Mexicans from different backgrounds are bravely investigating fuel trafficking at the ground level and helping understand this phenomenon. These include local journalists, another vital source of information given the valuable work they carry out in places like Guanajuato and Puebla. Risk consultants, advising potential energy investors, were also important. Many former security and intelligence officials are offering their services in risk consulting, bringing their expertise along with them.

My fieldwork experience enhanced my initial perception: fuel trafficking was a massive criminal market and its impacts were significant. Still, challenges remained: The secrecy surrounding criminal enterprises was worsened by the lack of commitment of Mexican authorities for transparency during the Enrique Peña Nieto (EPN) presidency (2012–2018). In January 2018, PEMEX classified fuel trafficking information as confidential, arguing that the security of the company's operations, personnel and competitiveness were threatened by making any information on the issue public (González 2018). These arguments were problematic, considering that the greatest threat to PEMEX is the black market itself. These claims also appeared dubious considering the accusations of high-level collusion and of involvement of PEMEX workers in fuel trafficking.

During the first stage of my investigation a report elaborated for Mexico's energy regulator, accessed by a transparency petition, represented a turning-point. Detailed data on the issue was finally revealed, further pointing to the importance of this criminal market and corroborating my interviewees' information. A second breakthrough was soon to come: In December 2018 the newly installed Obrador presidency announced a crackdown on fuel trafficking, presenting data that allowed to put together a comprehensive picture of Mexican fuel trafficking. Presidential conferences would then become a valuable source of information. This official data allowed to cross-check my informants' arguments.

It thus became clear that fuel trafficking consolidated as a paramount illegal enterprise and a priority for Mexico's criminal actors. My investigation now focused on finding explanations as to how this happened.

As such, this investigation contributes to the understanding of Mexico's deteriorating security situation and the recent history of its criminal protagonists. It also helps understand the dynamics pushing black-market diversification and clarify the mechanisms implemented by the mercurial criminal groups now dominant in the country.

I expect this to be one of the first academic investigations on fuel trafficking, a black market that will play a significant role in Mexico's security situation for years to come. Understanding its origins and the factors that made it flourish will prove of relevance for anyone who is interested in energy and security, illicit markets, criminality, modern-day Mexico and the challenges posed by the country's security crisis.

About This Research

Researching black markets poses many challenges. Accessing truthful information is problematic. As Hobbs and Antonopoulos (2014) argue, efforts to assess black markets are *"burdened by the inherent difficulty of estimating (or making estimations about) markets that are illegal and hidden"* (6). Mexican fuel trafficking is no exception. Inevitably, some ambiguities will end up in these pages, but this should not stop researchers from seeking knowledge on criminal phenomena and contributing to the field. As Peter Andreas (2010) claims: *"The point here is not that we should throw up our hands in frustration and conclude that the illicit world should be ignored and cannot be studied simply because it cannot adequately be quantified. Doing so would be the equivalent of the drunkard looking for his keys under the light-post because it is the only place he can see"* (45).

This research will be organised as follows: First, we will explore international cases of fuel smuggling to identify guidelines to understand Mexican fuel trafficking. To define these guidelines, we will develop: (1) the global presence of fuel-theft and trafficking, a black-market found in the Americas, the Middle East, Asia and Europe. Subsequently, a series of global patterns of fuel trafficking will be outlined including (2) the all-encompassing participation of diverse criminal networks in illegal fuel smuggling. As we will see, these criminal participants vary in their capacities and the sophistication of their operations. Afterwards, we will investigate another common characteristic relevant for this analysis: (3) the participation of actors linking the legal and the illicit realms that are key for the viability of fuel black markets. Finally, our last point linked to international cases of fuel trafficking will be (4) illegality camouflaging. Illegality camouflaging offers a protective layer which is not feasible in other criminal enterprises. Furthermore, camouflaging illegality allows criminal actors to occupy legal realms, offering many incentives.

Second, after the concepts derived from international cases, some theoretical concepts to examine the rise of Mexican fuel trafficking will be developed. These concepts will go from the regional to the national and are closer, but not exclusive, to the Mexican case. The first includes fragmentation and black-market diversification of illicit groups, a trend experienced in Mexico and Latin America. The second concept is the illicit co-option of different state institutions, particularly the state-owned enterprise (SOE) PEMEX.

To investigate the MFBM, we will explore the social organization of different black markets using economic sociology. As aforementioned, this field is of special interest to this research for its emphasis on the social organization of illicit markets, the actors involved in them, their social contexts and the characteristics defining them. Afterwards, an introduction to the MFBM will be made. Here, one of our key theoretical concepts for this research is further developed: the co-option of PEMEX by different state and non-state actors. As we shall see, this co-option phenomenon is crucial to the explosive expansion of the MFBM.

Beyond the Mexican context, these concepts should invite other researchers to apply them to other cases of fuel smuggling and use them for scholarly contributions. In order to illustrate this, prior to delving into the Mexican context we will use this research's theoretical framework and analytical tools to examine a fuel trafficking case that occurred between Libya, Malta and Europe. The objective of this section is to show this research's comparative value.

Third, after having introduced the theoretical framework combining international, regional and national-level concepts, we will delve into empirical data on the MFBM. We begin this analysis with the context that gave rise to the MFBM. This entails analysing Mexico's recent energy and security situation. This will be followed by a historical account of the emergence of the MFBM. Part of this section involves pinpointing the moment when the involvement of large criminal networks in this illicit enterprise was first detected. Identifying this is critical as it allows us to understand how the MFBM spread across Mexico and to explain how criminal networks became relevant participants. This analysis sets the stage for delving into the relevance of the MFBM. Here, multiple crosschecked sources are used to the explain the size of this black market, its profits and its operations. It is important to mention that official sources show inconsistencies. Despite their limitations, these are the best sources of data available, and they shine light on this criminal enterprise.

Fourth, after explaining the recent history and current situation of the MFBM, we continue describing the factors behind its explosive growth. The list is not exhaustive, but it includes the most relevant elements identified by my investigation and interviewees. These factors are: (1) the fragmentation of criminal networks and black-market diversification, (2) the co-option of grey actors and (3) the increase of fuel prices in Mexico between 2011 and 2018. These factors, though applied to Mexico, can prove relevant to other scenarios of organized crime illustrating the connection between black markets, institutional fragility and economic changes.

Fifth, after having analysed international and national cases, we take our analysis to the local level by examining two MFBM hotspots: Puebla and Guanajuato. Both cases are useful to illustrate how the theoretical concepts developed in this investigation translate into local contexts (local grey actors, state-level criminal fragmentation, black-market diversification, co-option of local actors and institutions). This will be also an opportunity to explore how the MFBM is influenced by local conditions. Here we will draw our attention to two local phenomena: community involvement in Puebla and criminal conflict in Guanajuato.

Consequently, the research will follow the subsequent structure (Illustration 1.1).

The final part of this investigation makes an update on the state of the MFBM after a large-scale crackdown during 2019 under the Obrador presidency. For this conclusive chapter we leave behind the original structure of the investigation (going from the macro to the micro) to return to a national-level analysis. The reason behind this is the possibility of putting the framework of this research to the test of the extraordinary circumstances brought on by the state clampdown. This section will delve into the energy and security context under Obrador as well as the crackdown strategy on the MFBM and its consequences.

Having gone through the book's structure I would like to mention some aspirations for this research. It is my hope that this investigation contributes to shed light on the importance of fuel trafficking as a criminal enterprise and inspires further investigations. The academic study of

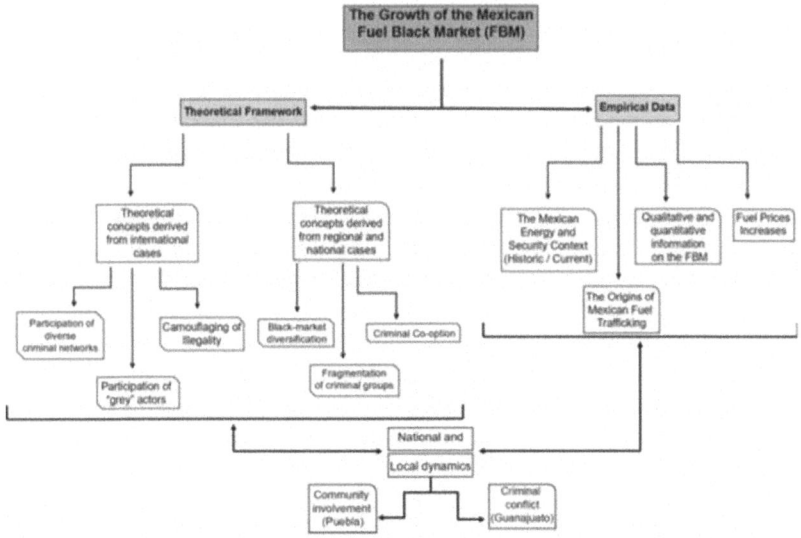

Illustration 1.1 Research theoretical framework and analytical tools

this criminal phenomenon still requires more work, and the field can only benefit from other comprehensive case studies. Further research is needed to clarify the causes, scope and effects of this black-market in other contexts and elucidate its differences and similarities at an international level.

I hope that, on a broader level, this investigation contributes with some insights on the diversification and sophistication of criminality in Mexico, Latin America and the world. A lot of attention is still centred around a US-promoted "drug war", when in many Latin American countries vulnerable populations are subjected to a diversified and violent criminal landscape, facing trauma and loss as part of their daily lives. Besides the undeniable American influence, many of the forces behind this situation are local and need academic attention.

I expect that this research helps contribute to the understanding of new profitable black markets operating worldwide. We need to better understand the contexts, actors and conditions that lead to this diversification and find answers to tend to its negative effects (particularly where their appearance leads to escalating homicidal violence). Despite these bold aspirations, I conceive this work as one contribution to a larger academic debate to inform alternatives regarding criminality, black markets and public security. The illicit world is shrouded by secrecy and the threat of violence, and as researchers in this field we must embrace and understand that definitive conclusions will remain stubbornly elusive. Having said this, those limitations should not dissuade us from striving to understand these pressing matters.

A Mexican Case Study, with Far Reaching Implications

This research is a Mexican case study. Yet its implications make it more far-reaching than what that description suggests. By developing the growth and consolidation of fuel trafficking as one of Mexico's main criminal markets, this research contributes to understanding the Mexican security crisis, the actors and forces propelling its violence and the trend towards increasing diversification that has transcended the so-called "war on drugs". Academics like Lessing (2015) have contributed to the interpretation of the Mexican security situation beyond the outdated "war on narcotraffickers". Yet this discourse is still pervasive beyond academic circles, and this research provides a more up-to-date interpretation with a timely analysis of the MFBM.

The implications of what is depicted in these pages are not restricted to Mexico. A case that is developed in this investigation involves the criminal dynamics between Mexico and the US and shows that fuel trafficking can transcend national borders, becoming a regional security challenge. This research provides an academic analysis on the issue of criminal diversification in Latin America, where the dispersion of illegal markets towards natural resource extraction has expanded (as in Africa and Southeast Asia). Some examples include the illegal exploitation of coltan in Colombia (González Garzón 2015) and the timber black markets of Brazil, Indonesia and the Democratic Republic of Congo (Boekhout van Solinge 2014). Consequently, the theoretical frameworks and arguments developed in this research can help researchers analyse fuel black markets observed elsewhere in the world (Ralby 2017), or other criminal enterprises involving illegality camouflaging and the participation of grey actors.

This research also shows how the realms of legality and illegality operate in tandem rather than separately (Garay-Salamanca and Salcedo-Albarán 2016). Still, official discourses conceive criminality and legality apart from one another, in constant confrontation. Yet this is hardly the case. Legal and illegal agents and entities routinely operate in collaboration, even though in other moments they may confront one another. This investigation provides evidence of this complex relation and argues against clear-cut separations.

During the years I researched and wrote about this subject, a recurrent misinterpretation kept appearing in different media outlets: the idea that crippling underdevelopment is inexorably linked to thriving criminal enterprises and criminal violence. It is my hope that this research instead shows that the presence of large-scale criminal networks and consolidated illicit enterprises is highly dependent on growth, infrastructure and many elements that could make a nation fall in the category of "developed".

Finally, I would like to make the point that this is one of the first scholarly treatments of the issue of fuel trafficking in any language. Personally, I hope that this investigation raises various questions and promotes further research on this criminal market that has been documented worldwide. This global diversity brings us to this research's first chapter.

Bibliography

"¿Qué Pasó el 3 de mayo en Palmarito, Puebla?" *Noticieros Televisa*, May 11, 2017. https://noticieros.televisa.com/ultimas-noticias/que-paso-3-mayo-palmarito-puebla/. Date accessed: July 13, 2021.

Andreas, Peter. 2010. "The Politics of Measuring Illicit Flows and Policy Effectiveness." In *Sex, Drugs, and Body Counts: The Politics of Numbers in Global Crime and Conflict*, edited by Peter Andreas and Kelly M. Greenhill, 23–45. Ithaca: Cornell University Press.

Atuesta, Laura, and Yocelyn Pérez-Dávila. 2018. "Fragmentation and Cooperation: The Evolution of Organized Crime in Mexico." *Trends in Organized Crime* 21: 235–261.

Ayling, Julie. 2009. "Criminal Organizations and Resilience." *International Journal of Law, Crime and Justice* 37: 182–196.

Beckert, Jens, and Matías Dewey. 2017. "The Social Organization of Illegal Markets." In *The Architecture of Illegal Markets: Towards an Economic Sociology of Illegality in the Economy*, edited by Jens Beckert and Matías Dewey, 1–38. Oxford: Oxford University Press.

Beckert, Jens, and Frank Wehinger. 2013. "In The Shadow: Illegal Markets and Economic Sociology." *Socio-Economic Review* 11 (1) (January): 5–30.

Beckert, Jens, Nina Engwicht, Annette Hübschle, and Matías Dewey. 2015. "The Black-Market as a Gray Zone." *MaxPlanckResearch*: 70–76.

Boekhout van Solinge, Tim. 2014. "The Illegal Exploitation of Natural Resources." In *The Oxford Handbook of Organized Crime*, edited by Letizia Paoli, 500–526. Oxford: Oxford University Press.

Bright, David A., and Jordan J. Delaney. 2013. "Evolution of a Drug Trafficking Network: Mapping Changes in Network Structure and Function Across Time." *Global Crime* 14 (2–3): 238–260.

Correa-Cabrera, Guadalupe. 2017. *Los Zetas Inc.: Criminal Corporations, Energy, and Civil War in Mexico*, 1–341. Austin: University of Texas Press.

Durán-Martínez, Angélica. 2018. *The Politics of Drug Violence: Criminals, Cops and Politicians in Colombia and Mexico*, 1–299. Oxford: Oxford University Press.

Esparza, David Pérez, and Helden de Paz Mancera. 2017. "Los huachicoleros: cinco lecciones para Enrique Peña Nieto." *Nexos*, May 15. https://redaccion.nexos.com.mx/?p=8102. Date accessed: July 13, 2021.

Garay-Salamanca, Luis Jorge, and Eduardo Salcedo-Albarán. 2016. *Macro-Criminalidad: Complejidad y Resiliencia de las Redes Criminales*, 1–191. Bloomington: iUniverse.

Garay-Salamanca, Luis Jorge, Eduardo Salcedo-Albarán, Macías Fernández Guillermo, Diana Santos Cubides, and Nathalia Guerra Villamizar. 2018. *Macro-Corruption and Institutional Co-optation: The "Lava Jato" Criminal Network*, 13–209. Bogota: Vortex Foundation.

Garzón, Juan Carlos. 2014. "From Drug Cartels to Predatory Micro Networks: The 'New' Face of Organized Crime in Latin America." In *Reconceptualizing Security in the Western Hemisphere in the 21st Century*, edited by Bruce M. Bagley, Jonathan D. Rosen, and Hanna Kassab, 117–131. Lanham: Rowman & Littlefield.

González, Nayeli. 2018. "Reservan 5 años daños de ordeña; Pemex alega "seguridad nacional"." *Excélsior*, January 1. https://www.excelsior.com.mx/nacional/2018/01/12/1213194. Date accessed: July 13, 2021.

González Garzón, Hermann David. 2015. "Transgresión de derechos humanos a raíz del tráfico ilegal de coltán en el Departamento del Guainía." *Revista Científica de la Escuela de Postgrados de la Fuerza Aérea Colombiana* 10: 151–168.

Guerrero, Eduardo. 2018. "La segunda ola de violencia." *Nexos*. April 1, 2018. https://www.nexos.com.mx/?p=36947. Date accessed: July 14, 2021.

Hobbs, Dick, and Georgios A. Antonopoulos. 2014. "How to Research Organized Crime." In *The Oxford Handbook of Organized Crime*, edited by Letizia Paoli, 1–27. Oxford: Oxford University Press.

Hope, Alejandro. 2017. "En Tiempos de Peña Nieto." *Nexos*, January 1. https://www.nexos.com.mx/?p=30852. Date accessed: July 13, 2021.

Hübschle, Annette. 2017. "Contested Illegality: Processing the Trade Prohibition of Rhino Horn." In *The Architecture of Illegal Markets: Towards an*

Economic Sociology of Illegality in the Economy, edited by Jens Beckert and Matías Dewey, 1–22. Oxford: Oxford University Press.

Kenney, Michael. 2007. "The Architecture of Drug Trafficking: Network Forms of Organisation in the Colombian Cocaine Trade." *Global Crime* 8 (3) (August): 233–259.

Lessing, Benjamin. 2015. "Logics of Violence in Criminal War." *Journal of Conflict Resolution* 59 (8) (December): 1486–1516.

Mastretta, Sergio. 2017. "Escenas del Huachicol Poblano." *Nexos*, May 12. https://www.nexos.com.mx/?p=32348. Date accessed: July 13, 2021.

OECD. 2016. *Illicit Trade: Converging Criminal Networks. OECD Reviews of Risk Management Policies*, 1–266. Paris: OECD Publishing.

Pérez, Ana Lilia. 2011. *El Cártel Negro: Cómo el crimen organizado se ha apoderado de Pemex*, 5–221. México: Grijalbo.

Pérez, Ana Lilia. 2017. *PEMEX RIP: Vida y Muerte de la Principal Empresa Mexicana*, 9–395. México: Grijalbo.

Petróleos Mexicanos. 2018. "Informe Anual 2017," 4–141.

Ralby, Ian M. 2017. "Downstream Oil Theft: Global Modalities, Trends and Remedies," 1–117. Atlantic Council Global Energy Centre, January.

Van Audenhove, L. 2017. "Expert and Elite Interviews in the Social Sciences."

Williams, Phil. 2001. "Transnational Criminal Networks." In *The Future of Terror, Crime, and Militancy*, edited by John Arquilla and David Ronfeldt, 61–97. Santa Monica: RAND Corporation.

Yashar, Deborah J. 2018. *Homicidal Ecologies: Illicit Economies and Complicit States in Latin America*, 1–368. Cambridge: Cambridge University Press.

CHAPTER 2

An Introduction to Fuel Trafficking

A GLOBAL INTRODUCTION TO FUEL TRAFFICKING

Smuggling hydrocarbons is a global criminal enterprise whose presence is not limited by national borders or cultural identities. There are documented cases in Asia, the Middle East, Africa, Europe and North and Latin America. Exploring some of these cases will help us distinguish between the various causes, similarities and differences that make fuel trafficking a global black market. But most importantly, they will allow us to define common threads behind fuel trafficking that will be crucial to understand the MFBM.

Researching and Defining Fuel Trafficking

Fuel trafficking is a diverse criminal activity. It involves small to large criminal networks and many factors can incentivize it, from cross-border national energy subsidies, difference in national fuel prices, taxation, conflict, livelihood loss and so forth. It affects countries with strong and weak institutions, impacting security differently. Therefore, a broad definition like the one elaborated by Happy (2015) is useful: a black market encompassing the theft of crude oil and its derivative products through diverse mechanisms (1). Another valuable definition, adding the elements of groups or individuals involved and the damage caused to the state, is the one proposed by Willis (2014): "*oil theft involves illegal siphoning of crude oil from oil facilities by individuals and groups at detriment of the state development"* (69).

So far, the phenomenon of fuel-theft and trafficking has been ignored by academia, with some exceptions. Some examples include Ikelegbe (2005), Burdin (2009) and Odalonu Happy (2015) who researched Nigerian oil trafficking. Kiourktsoglou and Coutroubis (2015) contributed by analysing the fuel trafficking operations of the Islamic State (IS). In 2005 while investigating the Nigerian case, Augustine Ikelegbe noted that oil trafficking grew from an amateur criminal activity *"to a very sophisticated industry which uses advanced technologies"* and that the *"stealing and smuggling of crude has become very extensive and large-scale since the late 1990s"* (Ikelegbe 2005, 221–222). More recently Ralby (2017) affirmed that trafficking refined products gained importance because derivatives are better protected against oil's price volatility (4). Ralby also warned that more rigorous academic research was needed.

Many of the people investigating fuel trafficking have been journalists, security analysts, policy advisers and risk consultants. In Mexico, the phenomenon has been analysed focusing on a fuel trafficking operation that took place in Tamaulipas in the mid-2000s. One such example is the journalistic investigation by Ana Lilia Pérez (2011) that revealed the involvement of criminal organizations in the trafficking of gas condensate. Another example is the work done by Sullivan and Elkus (2011). Despite these efforts more comprehensive and up-to-date academic research is still to be elaborated.

Fuel Trafficking: A Global Criminal Market

Stealing crude oil and its derivatives is achieved through diverse methods: pipeline extraction, tanker truck theft, theft in refinery and distribution terminals and marine piracy. Fuel black markets also vary across nations. In Nigeria, smuggling oil derivatives constitutes a large-scale market that generates profits estimated between $3 and $8 billion U.S. dollars (USD) a year (Ralby 2017, 15). Fuel trafficking in this African power constitutes a pressing threat to its national security.

In Colombia and Turkey, fuel trafficking became an important source of revenues for insurgent groups and the IS. Turkish and Colombian fuel smuggling represents a strategic threat to security, though to a lesser extent than the Nigerian case. In the European Union (EU), fuel smuggling is a profitable criminal enterprise that does not represent a substantial risk to the continental bloc.

Fuel trafficking is a global phenomenon that affects countries in different regions, nations with distinct institutional capacities and with diverse energy sectors (consuming and producing countries, refining-specialized nations and energy transport hubs). These cases show variations and similarities. The following global cases will work as examples to introduce some theoretical points that will be used to analyse the Mexican case. They are also intended to show, concisely, the diversity of fuel smuggling in hopes that they can help inspire further research.

Having clarified the intention of such cases, a repeating pattern is that fuel smuggling (national and transnational) can be triggered by energy price disparities caused by subsidies and taxes.

Of Fuel Trafficking, Taxes and Subsidies

According to international cases, any policies that produce fuel price disparities (subsidies or taxes) are likely to lead to the formation of fuel black markets, national or transnational. Examples of such dynamics have been detected in Africa, Asia, Europe and Latin America (Ralby 2017, 87).

Fuel trafficking between Thailand and Malaysia exploits price disparities between these two nations. These differences are a consequence of subsidies. According to Thai authorities, in February 2015 unleaded high-octane gasoline (ULG 95) cost $0.55 USD/litre in Malaysia, while in Thailand it cost almost double ($1.08 USD/litre). By August 2016, the price of ULG 95 in Malaysia was $0.47 USD/litre, while in Thailand it was $0.92 USD/litre. These price differences generated incentives for criminal actors to profit through transnational smuggling.

In Turkey fuel trafficking takes advantage of internal price disparities while transnational smuggling exploits regional instability. Energy prices within Turkey are higher in the north than in the south. This generated the conditions for a black market reliant on adulterating fuel formulas and artisan refining. One example is *Number 10 Oil*, a low-quality artisanal diesel that is sold across the country.

The illegal fuel market in the EU has found favourable conditions to consolidate because of the price disparities amongst the bloc's countries. Different criminal actors take advantage of tariff and taxation differences to commercialize in European nations where energy prices are higher.

One example is fuel trafficking between the Republic of Ireland, Northern Ireland and Great Britain. Most fuel goes from Poland to Ireland. In March 2016, Polish diesel cost €0.91 while in Ireland it cost €1.10 (McAleese 2016).

In Colombia, the price difference with neighbouring Venezuela catalysed a thriving fuels black market. Once in Colombia, the price of the gallon of Venezuelan gasoline increases by 3,700 times its original value (Villalba 2018). Fuel trafficking routes between these two nations show a price increase dynamic like that of narcotics: as the product gets further away from its place of origin, its value increases exponentially. Within Venezuela, the price of gasoline grows 700 times (from $0.00055 USD to $0.040 USD) between Maracaibo y Paraguaipoa, a town on the border with Colombia. Once the fuel crosses into Colombia prices increase between 145,354.5% and 558,081.8% (from $0.80 to $3.07 USD).

Tariffs and taxes can be catalysts for the emergence of illegal fuel markets, national or transnational. Yet this phenomenon will not be further explored in this research as it is not particularly applicable to the Mexican context. I include it as a global trend of fuel trafficking that could be the subject of further research. To determine the points relevant to this investigation, we must explore other international fuel trafficking cases.

Criminal Networks of All Sizes Participate in Fuel Trafficking

Fuel trafficking is a diversified illicit activity in which criminal groups of all capacities participate. In Nigeria, a large criminal/insurgent network known as the Niger Delta Avengers (NDA) is greatly involved in fuel smuggling in the southern regions of the country. The operation capacity of this group allowed it to decrease Nigeria's oil production by more than 36% through sabotages in 2016 (Ralby 2017, 24) and it represents a threat to the integrity of the state because of its links to separatist movements.

On the other side of the world, highly diversified criminal networks participate in fuel trafficking operations between Thailand and Malaysia, profiting from transnational price disparities. In 2013 a government crackdown dismantled a network involved in hydrocarbons smuggling, human trafficking, narcotics and car theft operating in Thailand and its neighbouring countries ("Police Chief Says Oil…" 2013).

Turkey, a transport hub playing a crucial role for Europe's energy supply from the Middle East, has also experienced fuel trafficking. Estimates claim that fuel smuggling into Turkey grew as much as 300% since the outset of the Syrian civil war (Gingeras 2014). The product destined for Turkey is refined in Syria and converted into a low-quality diesel used to power generators. Most of the smuggling uses jerry cans transported on horses through rough terrain or by boats. Irrigation pipes crossing the border into Turkey illegally transporting fuel have also been discovered.

The IS was involved in fuel trafficking[1] and managed large-scale smuggling from Syria into Turkey using truck convoys. Smuggled IS shipments were promptly transferred into tankers departing from the Turkish port of Ceyhan (Kiourktsoglou and Coutroubis 2015). Brokers were used by IS as middlemen, indicating that many of the agents involved in the operation did not know the origin of the fuel and revealing a cell structure in which *"lower levels of communication increased security for the organization"* (Vigil 2016).

In the EU, different criminal actors actively participate in fuel trafficking operations, from small traffickers to large illicit networks. In March 2016, a Polish criminal network that trafficked adulterated diesel mixed with oil derivatives from Russia and Belarus was discovered. The group commercialized this product as biodiesel within the EU. One example of larger groups operating in Europe was found in 2017, when a transnational operation of fuel trafficking was uncovered between Libya, Malta and Italy. This case will be analysed at the end of this chapter. It is important to mention that there are more documented cases of fuel-theft in Europe.[2]

These operations cannot function with actors that are solely illegal. This is where an important type of agent, who connects the criminal with legal realms, comes into play.

---

[1] According to the US Treasury Department, IS produced approximately 50,000 barrels per day (bpd) and generating up to $40 million USD a month from its trafficking.

[2] Two more cases that exemplify alleged fuel-theft and trafficking in Europe are: the MiRo refinery in Germany (Bloomberg 2013) and the state-owned rail company BDZ in Bulgaria (Novinite 2015).

Grey Actors: An Unescapable Element in Fuel Trafficking

Global fuel trafficking prospers through the participation of "grey actors" who have presence between "black" illegal markets and "white" legal markets. These agents can be politicians, government officials, law enforcement agents, bankers or entrepreneurs that provide criminal networks with strategic resources for their operations. Some of these resources include money-laundering, guaranteeing impunity, co-opting authorities and establishing long-term agreements (Garay-Salamanca and Salcedo-Albarán 2016, 33). Grey actors fulfil the crucial role of inserting criminality into the legal workings of state and society, while providing inputs of strategic importance such as information, technical know-how and distribution channels that allow fuel trafficking to thrive. These actors can hold legitimate positions while their activities challenge the law (Yashar 2018, 70). The involvement of realm-linking grey actors, particularly those belonging to the state, can compromise entire institutions (ibid.). This reality questions the nature of the state, which can incorporate both the licit and illicit, assuming the role of lawful authority while also engaging in criminal operations.

Mayntz (2017) reinforces this grey actor thesis by identifying criminal enterprise agents that *"engage both in legal and illegal actions, who are moving between two worlds, acting legally and then again illegally"* (8). These actors can create *"interfaces"* that are *"boundary-spanning"* and connect legality to illegality. What this means at a larger level is that the existence of *"interfaces between legality and illegality... alerts us not so much to the dark side of the social world as to the many shades of grey that lie between black and white"* (9). Legality and unlawfulness do not operate in isolation but rather coexist closely, unlike what hegemonic discourses claim.

Grey actors enable illicit fuel enterprises to have a transnational scope. In Nigeria, networks traffic fuels to the Gulf of Guinea region and have reached Australia, Eastern Europe, France, Greece, Morocco, Netherlands, Russia and Venezuela. In a 2009 case, resources generated by Nigerian fuel trafficking were traced to Ivory Coast and Senegal, to then be channelled through the French financial sector to bank accounts in Syria and Lebanon (Burdin 2009, 5). Nigerian authorities play a key role in facilitating documentation and logistical support for this black market (Ralby 2017, 17).

Fuel trafficking in Thailand, for example, has thrived through the involvement of high-level officials and members of the royal family. In November 2014, security officials were detained for their involvement in fuel trafficking. They were arrested for a series of crimes, including receiving a bribe of $4 million USD from fuel smugglers between 2012 and 2014. Another high-level detainee was lieutenant general Pongpat Chayaphan, a senior security official ("One More Month in Jail..." 2018). Chayaphan had family members belonging to the royal family who were also implicated in fuel trafficking. Scandal was such that the wife of the Thai crown prince and her family were stripped of their royal status (Ralby 2017, 57). Just as in Nigeria and Thailand, fuel smuggling also involved state representatives in Turkey. In 2013 authorities arrested the sons of various ministers for bribery, theft and trafficking. Investigations uncovered that the detained were part of a criminal network that exchanged gold for Irani gas and oil (Gingeras Ryan 2014).

Fuel trafficking in the EU also takes place with the participation of grey actors. In 2017, a fuel trafficking network operating between Libya, Malta and Sicily was uncovered. This network included businessmen dedicated to maritime trade, with alleged mafia connections. The group's fuel trafficking operations began after the collapse of the Gaddafi regime in Libya in 2011. Since then, an armed militia gained control over a refinery that included a marine terminal. This group supplied fuel, transporting it using tanker ships sailing to international waters. Once out of any state jurisdiction, the tankers met ships owned by the entrepreneurs to hand in the hydrocarbons in a maritime transfer between vessels known as "bunkering". The businessmen used their companies as facades for transport and distribution. Subsequently, the ships took their cargo into Malta for storage and distribution in Europe. Maltese authorities validated the documentation they presented as lawful (*Times of Malta* 2018). This case, which will be used to apply this research's theoretical framework in this chapter, illustrates the diversity of actors involved in fuel trafficking.

Fuel trafficking is also present in Latin America. In Colombia, the insurgent group *Autodefensas Unidas de Colombia* (AUC) has been involved in fuel trafficking with the support of grey actors. It was revealed that the AUC was stealing large amounts of crude oil in collusion with workers of Ecopetrol, Colombia's state oil company. Between 2001 and 2003 it is estimated that the AUC stole $10 million USD from various pipelines in Colombia. The AUC also imposed tariffs on tanker trucks transiting through its controlled territories. According to an AUC leader,

Ecopetrol allowed the group to steal oil from its northern pipelines in exchange for protecting the company's infrastructure from other criminals (Pachico 2011). Crude oil is important for criminal agents in Colombia since it can be turned into gasoline through artisanal refining. The gasoline is then used as a precursor for cocaine production ("Venezuela: A Mafia State?" 2018, 49).

In the Colombian-Venezuelan border, a criminal network known as the Contraband Cartel controls a fuel black market estimated to represent 70% of the economy of the border city of Macaio, a distribution node for fuel smuggled from Venezuela receiving roughly 434,750 litres of gasoline daily (Villalba 2018). This criminal group also employs grey actors for its trafficking operations: once in Colombia, the group relies on cooperatives legally entitled to commercialize fuel for distribution, camouflaging their illicit derivatives as lawful. The Contraband Cartel distributes 117,500 gallons of fuel monthly, generating profits of $7.5 million USD (Pachico 2011). In Venezuela fuel trafficking also thrives with the participation of grey actors. Photographs have shown the Bolivarian National Guard participating in transborder fuel trafficking ("Venezuela: A Mafia State?" 2018, 49).

Regardless of the region, culture or peculiarities of each nation, grey actors play a leading role in fuel trafficking. They are an inescapably crucial element in this black market, allowing the operations of this criminal enterprise to consolidate. This drives us to the next common denominator in fuel trafficking: the possibility of illegality camouflaging.

Fuel Trafficking: Camouflaging Illegality

A pattern in more sophisticated versions of fuel-theft is camouflaging illegality, or making it appear legitimate. This is implemented by forging or obtaining documents and altering modes of transport and commercialization. Fuel trafficking allows for this camouflaged illegality since its products are not inherently illegal. It also makes operations harder to identify, minimizing exposure: for example, a tanker truck delivering stolen fuel in a gas station could hide its illegality in plain sight. This camouflaging allows criminals to deploy "*invisible supply chains*" (Ralby et al. 2019) that may be fully illicit (from extraction to commercialization) or invade the legal market in different segments.

In Nigeria, ships transport stolen fuel using bills of lading detailing the type, quantity, and destination of their cargo. These documents are issued by custom agents, administrative authorities and on-site officials (Ralby 2017, 18). Camouflaged illegality is one of the main contributions of grey actors to criminal networks who profit from fuel trafficking, enabling access to documentation that simulates "lawful" operations. But camouflaging can transcend paperwork.

Fake labelling and the adulteration of fuel formulas can complement forged documentation and invoices. In the EU, fuel adulteration is a common practice amongst sophisticated criminal networks. Various criminal groups in the EU label high-priced fuels as derivatives subject to lower tax burdens, a crime that is known as "hydrocarbons fraud". Another option is tampering with formulas to imitate other derivates *"assessed for lower tariffs or VATs, and then selling them as higher-taxed fuels"* (ibid., 71).

Illegality camouflaging also applies to transport. In Thailand, fishing vessels are modified to transport hidden illegal fuel in bulk (Thepbamrung 2013). Unlike other products like narcotics, fuel gives criminals the possibility of camouflaging their operations while taking place in the open. These three traits (criminal networks participation, the importance of grey actors and illegality camouflaging) will be relevant to understand the Mexican fuel trafficking phenomenon.

Having explored the research and definitions of global fuel trafficking and given an overview of its diversity and common traits, we may return to Mexico. Another set of theoretical concepts instrumental to understanding Mexican fuel trafficking thus arise.

THEORIZING MEXICAN FUEL TRAFFICKING

Diversification and Criminal Fragmentation

As the war on drugs intensified in the region, Latin American criminal networks added new illegal enterprises to their portfolios beyond narcotics. This diversification has been defined as an increasing participation in multiple criminal activities, particularly local predatory enterprises like human trafficking, extortion, kidnapping, drug retail and *"theft and sale of gasoline"* (Garzón 2014, 124). These criminal markets can be as profitable as narcotics and offer other advantages to criminals. They allow access to profits while involving local populations and consolidating social support. Local black markets also offer possibilities for money-laundering, infiltrating the legal economy and exploiting lawlessness to corrupt authorities (ibid., **125**). Diversification leads to new revenue sources to sustain organizational continuity.

Explanations for criminal diversification can also be found in business literature, considering the profit-motive driving criminal enterprises. Business diversification can be defined as a firm's *"entry into a new market with which it is currently not associated"* (Park 2016, 6). From this perspective, increased competition amongst criminal networks is also reason for diversification. This can combine with the underuse of resources like smuggling routes, transport structures or criminal territorial domain. If profitable activities can be carried out in a territory under criminal control, not doing so would represent a wasted opportunity. The business side of criminal diversification is also further justified in an environment of increasing state repression and violent competition enhanced by fragmentation: diversifying into other markets can allow groups of all sizes to obtain resources intended *"to overcome the pressure exerted by Mexican authorities"* (Medel and Thoumi 2014, 14) and their growing number of criminal rivals.

This research argues that one of the factors behind the growth of fuel trafficking in Mexico between 2011 and 2018 is the fragmentation of criminal networks and their diversification into new illegal markets. The causation between the two is complex and reciprocal (fragmentation can lead to diversification and vice versa). Still, there is an academic consensus that fragmentation has compelled criminal groups of different capacities to branch out into markets beyond narcotics such as extortion, kidnapping, timber, human trafficking and fuel-theft (Felbab-Brown 2019, 12). This has driven criminal networks to develop specialized structures maintained through a vicious cycle of *"territorial disputes, violence, extraction of local resources, persecution, the need for more resources, more violence, more persecution"* (Fuerte Celis et al. 2018, 16).

Illicit organizations, in Mexico and the Americas, have shown to prioritize territorial control of subnational territories to guarantee their survival and the use of homicidal violence to achieve it. Territorial control entails not just spatial influence, but also control over government entities/representatives, information and social interactions. This control seeks to establish dominance over profitable black markets that can be transnational, national, regional or local in scope. The resources generated by these criminal enterprises are key for the organizations' sustainability. Potentially high profits act as incentive for organizations to compete for illicit rent capture, mitigate internal divisions or face a competing criminal threat. Fragmentation of criminal groups fuels organizational competition and diversification towards new black markets, which in turn brings increasing homicidal violence. Homicides have become the central means to *"defend and expand the territorial reach of these illicit territorial enclaves"* (Yashar 2018, 66).

Analysts have cited the kingpin strategy, implemented since the Calderón presidency (2006–2012), as the cause for criminal fragmentation (Felbab-Brown 2019), though not the only one. Internal group dynamics also play a role. Episodes of splits have been followed by cooperation amongst criminal networks pursuing self-preservation. Fragmentations may occur because of state repression, but also because of criminal competition. By analysing homicide data in sites with narco-messages, Atuesta and Pérez-Dávila (2018) found that in 2007 the Mexican criminal landscape had 5 groups participating in less than 20 violent events. By 2011, there were 80 groups involved in more than 1,000 events (237).

Having analysed these incidents, fragmentation amongst Mexican criminal networks can arise because: (1) existing alliances collapse, (2) a faction gains independence and (3) internal succession occurs. In these cases, fragmentations lead to disputes over territorial control and increased violence. Through violence, criminal groups achieve their goals, which are mainly economic revenues and organizational continuity (Fuerte Celis et al. 2018, 3). It is still worth mentioning that in most fragmentation cases the state played a role by eliminating criminal leaders. Prioritizing criminal leaderships has proven to be short-lived and counterproductive regarding violence in Mexico, yet it is an effective way for politicians and security institutions to show *"demonstrations of success"* (Felbab-Brown 2019, 15).

Alliances take place because (1) debilitated groups collaborate after a fragmentation, (2) confronting a shared enemy or (3) to expand territorially (ibid., 240). Criminal alliances are agreements between multiple criminal groups to oppose a common enemy, to avoid or endure violence within a territory or to pursue provisionally shared interests. Alliances for territorial conquest are relevant because territorial control allows criminals to profit from illegal markets. This applies to drug and human trafficking, and it is particularly relevant when the enterprise depends on the presence of energy installations. The MFBM requires control over territories that can be used for illegal extraction of different derivatives (in pipelines, refineries and storage and distribution terminals) *"as well as control over intermediary areas for storage and shipping"* (Fuerte Celis et al. 2018, 9).

A broken alliance between groups can trigger fragmentation and conflict once common interests disappear. Alliances do not have to materialize within groups that are structurally independent, and criminal networks can create specialized sub-networks to exploit certain black markets or fulfil specific objectives (establishing armed wings during turf wars). Associations with created sub-networks can prove as fragile as the ones between separate groups with autonomous structures and resources.

These categories do not necessarily operate isolated. For example, an alliance could form between smaller groups confronting a common rival. A fragmentation could follow, caused by the incarceration or death of a leader and/or groups acting autonomously. Criminal coalitions are a set of heterogeneous interests sharing a common goal. Once this goal vanishes, groups seek autonomy which can trigger fragmentation. These scenarios reflect that Mexico's criminal landscape is mercurial, with illicit actors in a process of constant change and adaptation. Diversification can be connected to expanding income sources when criminal networks find themselves under pressure to confront rivals and state forces, and have their capacities diminished to engage in transnational narcotics (Durán-Martínez 2018, 8–9).

As criminal competition increases so does violence to eliminate competitors, reinforce internal cohesion and establish control over resources and allegiances. Groups of different capacities have become more professionalized and embedded in different criminal economies. Therefore, the actors competing in Mexico for territorially embedded black markets have increased. Neighbouring geographies with lucrative criminal enterprises incentivize groups to migrate in the face of state crackdowns, in-fighting and violence from criminal competitors. Between 2007 and 2011 it is estimated that the presence of criminal groups increased in 20 Mexican metropolitan areas (of a total of 59) (Fuerte Celis et al. 2018, 9); indicating this territorial expansion trend linked to diversification that has consolidated in the country. These migrations are met with competing violence creating a vicious cycle. This local-level organizational competition for territorial control has led to increasing homicidal violence, which according to Yashar (2018) is the main reason behind this phenomenon (127).

Dynamics of fragmentation and cooperation have been key in the disappearance of vertical drug cartels and the rise of horizontal and resilient networks. Fragmentations and alliances have driven criminal groups towards new rivalries, allowed territorial expansion and made them transition towards more decentralized structures. This has given way to a more intricate criminal context whose actors are harder to define, to the point that *"it is almost impossible to follow each split, revenge or coalition between and within groups"* (Atuesta and Pérez-Dávila 2018, 259). Academics identified between 2007 and 2011 more than 200 groups, many formed for less than a year, while others changed brand-names and objectives (ibid.).

Mexico's modern criminal landscape is composed of groups of different origins, independent and diversified criminal factions cooperating to fulfil temporary goals. Groups of different capacities engage in diverse and profitable illicit markets, with small and medium groups becoming more relevant actors competing for profits. Some have become regional extending their operations across states and have shown the resources to compete against macro-criminal networks at a local level (as the Guanajuato case study exemplifies). To add intricacy to the situation, available official information has pointed out that small criminal cells are operating within this vastly diversified, competed and violent illicit underworld, engaging in different black markets and reaping vast profits. Extortion for example has become a widespread local level *"currency of control"* (Yashar 2018, 90).

In such scenarios, fragmentation becomes a catalyst for black-market diversification, as well as underused resources and competition. The Zetas left a distinct legacy of criminal enterprise diversification since the mid-2000s (Enciso 2017) that will be explored in chapter three. The dynamics of criminal fragmentation, combined with this innovation of this Tamaulipecan group, led to increased criminal involvement in other black markets, including fuel trafficking. As groups fragmented and the criminal landscape became more contested, diversification beyond drug trafficking became an imperative and fuel trafficking consolidated as one of the most profitable enterprises.

Black Markets: A Theoretical Overview

The growth of Mexican fuel trafficking is linked to the consolidation of the MFBM. Before developing the details of this criminal enterprise, we will explore some theoretical concepts behind illicit markets. For the case of this investigation, we will adhere to an economic sociology perspective because of the emphasis this field makes on the *"social organization of markets"* (Beckert and Dewey 2017, 2), addressing the *"the consequences of the illegal production, distribution, and consumption of illegal products for the architecture"* (ibid.) of black markets. This field of research proves of great value considering the interest of this investigation in defining the actors behind Mexican fuel trafficking and clarify their roles. Each black market reacts differently to the social context where it operates. Studying an illegal market should thus analyse the specific social phenomena surrounding it, such as mapping the actors involved, how secrecy is maintained and its operational methods.

A black market is the systematic exchange of goods and services for money under competitive conditions in which the products or their *"production exchange, or consumption violate legal stipulations"* (Beckert and Wehinger 2013, 7). Black markets are different from one another, some can be purely illegal in all their steps while others can be illegal activities that take place within lawful realms. Beckert and Dewey (2017, 4–5) propose these typologies:

1. Black markets in which offered goods or services are originally forbidden, with their trade and consumption are illegal. Some examples of these are narcotics or child pornography and are markets operating segregated from the legal economy.
2. Illegal markets where traded products are legal but have been acquired by the supplier unlawfully. In the cases where the buyer is aware of the products´ illicit origins, commercialization is considered illegal. These second type of black markets offer the possibility of trading the goods or services in isolated illegal markets or channelling them into legal markets; a relevant trait for this research.
3. Black markets where the products offered are counterfeited. Production might not be illegal, but trading is and, as the second type of unlawful markets mentioned above, transactions can occur in illegal markets or in legal spaces.
4. Products that are legal, but their trading is not (human organs). Market transactions in these products are morally offensive, making them operate separated from the legal economy.
5. Finally, illegal markets where the production, exchange, and consumption of products are legal, but violate regulations. These illegal markets are complex because their violations can take many

forms where the legal and illegal realms entangle constantly. Examples include violations of environmental regulations, product safety standards, trade licences or tax evasion.

Given how heterogenous black markets are, these definitions and their characteristics can overlap making illegal markets hybrid models. These typologies help us navigate the diverse landscape of black markets. Black markets can differ in their characteristics and present hybrid incarnations (which is the case with fuel illegal markets) (Table 2.1).

In the "Illegal Products (type 1)" markets, the illegality of transactions are visible while in the other markets (types 2 to 5), their illegality is *"much more covert and becomes invisible further down the value chain"* (ibid., 8). This concurs with the argument of illegality camouflaging in fuel trafficking and the deployment of invisible supply chains. As the table above shows, fuel trafficking's activities make the black market fall under the categories of "Stolen Products (type 2)", "Counterfeits (type 3)" and

Table 2.1 Illegal/Black market types

Type	Trading with illegal products	Consumption/ possession illegal	Market exchange illegal	Violation of regulation	Examples
1. Illegal Products	X	X	X	X	Narcotics
2. Stolen products		X	X	X	Stolen Antiques Fuel Trafficking
3. Counterfeits		(X) Illegal trade only	X	X	Luxury Item Fakes A Branch of Fuel Trafficking
4. Legal Products Illegally Traded			X	X	Human Organs
5. Violations of Regulations				X	Informality A Branch of Fuel Trafficking

Elaborated by author with information from Beckert and Dewey (2017)

"Violations of Regulations (type 5)". The international examples already explored depict how fuel smuggling can rely largely on stealing oil and derivatives (Nigeria) as well as cases where fuel formulas are adulterated (Turkey, the EU), and where regulations are violated.

Economic sociology not only concurs with the illegality camouflaging argument. According to Beckert and Dewey (2017), only in exceptional occasions black markets operate entirely within illicitness making them "*completely detached from legal exchanges in most parts of the value chain*" (16). Most of the times unlawful markets operate intertwined with their legal countertypes. Black markets have "*points of interpenetration where the illegal economy and the legal economy interact*" (ibid.) such as money-laundering. This makes a case for the co-option of grey actors linking legality and illegality: for criminal agents the close association with legal markets and organizations reduces the peril of detection, increases consumers and capital and allows to camouflage illegal actions (ibid.). This is further reinforced by the argument about the options criminal actors have to reduce transaction risks in illegal markets. Criminal actors can selectively pursue cooperation with determined state agents that benefit from the illicit enterprise while solving "*coordination problems*" and creating "*market stability*" (Beckert and Dewey 2017, 21). By co-opting state agents, criminal actors establish protection agreements that emulate barriers of entry and influence competition (ibid., 22).

An important characteristic of illegal markets is that, for structural reasons, they are predominantly determined by demand. This is because the supply side of most illegal markets cannot promote their goods and services amongst consumers. For a black market to consolidate it largely depends on the existence of demand. Illegal markets can also differ from one another because of the support and resources they can generate amongst different communities. In several cases the products or services of certain criminal enterprises can be repulsive (illegal narcotics, human trafficking). But in other cases, "*there are goods and services that provoke little rejection and may be met with tolerance, either because of their very nature, because they are embedded in tradition, or because they are considered vital for life*" (Beckert and Dewey 2017, 13).

Some communities' acceptance of certain black markets can even lead to scenarios of "*contested illegality*" (Hübschle 2017, 74) that legitimizes illicit activities. The reasons for this "*contested illegality*" to materialize are complex and will prove relevant for the case of community involvement in fuel trafficking in the state of Puebla developed in this investigation.

These are the theoretical contributions that economic ethnography makes to the interpretation of illegal markets, and which will help to understand the MFBM.

The Fuel Black Market and the Co-Option of PEMEX

The consolidation of the MFBM was largely achieved through the co-option of PEMEX by different criminal state and non-state actors. Co-opted State Reconfiguration (CStR) is defined as the actions of legal and illegal actors to alter the political and economic regime to systematically influence the formulation and application of laws, norms and policies, in order to obtain continuous benefits while validating their interests legally, politically and socially (Garay-Salamanca et al. 2018, 13–45).

In CStR agreements are established between state and non-state agents in the legal and criminal realms. Under this model of state reconfiguration, there is a scenario of instrumental capture of state institutions *"to reproduce and legitimate unlawful procedures and egoistic benefits"* (ibid., 27). This possibility of institutional assimilation drove researchers to coin the concept of Co-opted Institutional Reconfiguration (CItR). CStR and CItR are scenarios in which private and public legitimate actors establish agreements co-opting lawful and unlawful agents and vice versa. These agreements result in the coordination of shared interests.

For CItR to have viability, it requires grey actors providing key supplies (expertise, information, facilitating access, documentation) and concealing criminal actions. For the Brazilian embezzlement case *Operação Lava Jato*, Garay-Salamanca and Salcedo-Albarán (2018) developed the term *Systemic Macro-Corruption* given the complexity of implicated actors, the sophistication of the mechanisms they implement and the territorial and institutional scope of their actions (ibid., 30). A process of Macro-Corruption involves the coordinated participation of several public and/or private agents, including companies, criminal organizations and grey actors implementing various actions and agreements. Macro-corruption entails the manipulation of institutional norms and procedures, large money-laundering operations, establishing façade companies to profit and, most importantly, co-opt institutions and officials to generate stable agreements (political, entrepreneurial, bureaucratic and institutional). The agreements derived from this co-option generate sustained benefits for those involved with *"sophisticated forms of corruption beyond a single bribe"* (ibid., 31).

The networks behind the MFBM are diversified, and their institutional co-option of PEMEX fits within the definition of Systemic Macro-Corruption. These groups co-opted PEMEX through practices belonging to grand-corruption schemes including *"the use of fake receipts and falsified certificates"* (Rufyikiri 2016, 7). Through these measures criminal networks compromised a large part of PEMEX's effectiveness and diverted its institutional purposes. With the co-option of authorities and high-level officials these grand-corruption schemes can be maintained with impunity for sustained periods of time.

Other researchers describe similar scenarios to CStR and CItR. Durán-Martínez (2018) argues that, in Mexico's case, criminals do not pursue state control: they focus on local power struggles, explaining their violent behaviour and challenge to state sovereignty. She also contends that in criminal conflict situations the state cannot be understood as a unitary entity, it can be co-opted by circuits that can end up confronting one another. Durán-Martínez's framework coincides with the possibility of state institutions being selectively co-opted. Illicit organizations, therefore, should not be understood as polar opposites of the state. Rather, through co-option, they can be composed of both state and non-state actors. This is how criminal organizations *"assume both state and parastatal forms"* (Yashar 2018, 24).

Aretxaga (2003) goes further in affirming that the state as a unified structure does not exist, considering the contradictory ensemble of practices, processes, differences and power struggles within it. She also affirms that the state can lose ordering functions that make it a unitary force, like military and policing capacities. These functions in many cases get taken over by paramilitary and criminal groups (398) that use state persecutory power against state actors such as officials, politicians, military personnel, police, judges and prosecutors (402). Lessing's (2015) work also concurs with CStR and CItR, arguing that Mexican criminal networks deploy *"violent corruption"* against the state, targeting enforcement agents for co-option. This phenomenon entails the possibility of the state being co-opted by circuits and, as the case of MFBM will show, violent corruption not only targets law enforcers, but other state representatives.

The cases developed in this research show that the consolidation of the MFBM presents these characteristics and makes this a case CStR and CItR sustained through process of Systemic Macro-corruption. It is through these co-option processes that the MFBM expanded without state intervention, protected by impunity and by Macro-corruption agreements across PEMEX's hierarchy.

So far, we have explored this investigation's theoretical framework and analytical tools, based on examples from the international and Mexican context. And even though this investigation focuses on Mexico, this frameworks and tools have a comparative value that can help analyse cases of fuel trafficking (and illegal extractive enterprises) elsewhere. This is not fortuitous, different sources point to the value that single case studies have to elucidate wider-ranging phenomena. Siggelkow (2007) argues that the existence of phenomenon can be thoroughly described by single case studies (as cited in Gustafsson 2017, 3). Dyer and Wilkins (1991) maintain that single case studies produce careful research that allows a deeper understanding of the analysed subject and help craft higher quality theory (as cited in Gustafsson 2017, 3 and 4). Therefore, this single case study on fuel trafficking has comparative value beyond its setting. And this takes us to focus our attention, and this research's theoretical framework, on Libya, Malta and Europe.

The Maltese Connection

In 2017 Italian authorities detained several individuals for fuel smuggling. The network behind this ran a transnational operation that made over €30 million in profits (Vella 2017a), from at least 2014 to 2017. This case illustrates many elements of this research's argument and analytical tools (participation of diverse criminal networks, grey actors, illegality camouflaging, black-market diversification, illicit groups fragmentation and co-option) and shows how these tools can help analyse other international instances of fuel trafficking. It is also a case for readers who may be interested in fuel trafficking beyond Mexico. This example also shares similarities with Mexico's transnational Burgos Basin case, analysed on this research.

Before continuing, I would like to point out that I will not detail the fragmentation of illicit groups within this case. This point is a matter of specialized research by someone more knowledgeable about the highly complex situation of the Libyan civil conflict, where a myriad of armed groups constantly change identities and allegiances (Stephen 2014) and is well beyond my field of expertise.

Introduction

The video presented by the Italian *Guardia di Finanza* begins with a night vision shot of two vessels at sea. As they approach one another, both crews throw their moorings at each vessel, tying both ships together. The crew members suddenly pass from ship to ship two hose pipes ("#Catania. #Riciclaggio di gasolio" 2017). The vessels are bunkering illegally obtained fuel. What is behind these images is a criminal network that conducted a fuel trafficking operation that transcended multiple cultural backgrounds and jurisdictions, and that has its starting point in Libya.

The Libyan Security Context

Libya's criminal underworld changed dramatically after the demise of the Gaddafi regime. Beforehand, the nation's different black markets were regulated by the state, dividing these enterprises amongst different actors. After the collapse of the regime, fragmented violent contestation over the control of criminal markets followed. This translated into localized conflicts amongst individual actors, networks and communities, all competing for dominance of Libya's illicit economies and trafficking routes. The instability that followed 2011 accelerated the evolution of Libya's trafficking economies, where the dominance of informal goods was increasingly substituted by drugs, weapons, counterfeits, people and fuels (Eaton 2018, 8). Border communities became dependent on smuggling activities. This, as we shall see in this investigation, is a similarity with the MFBM.

Missing state enforcement allowed for the growth of trafficking networks, reaching *"industrialization"* levels of organization and integration (Eaton 2018, 8). One of the key phenomena that has reshaped Libya's illicit landscape is the appearance of multiple armed groups, promoting violence and fragmentation of illicit factions. Rent seeking of trafficking operations by armed groups and their direct involvement in them has become commonplace. In this context of an increasingly disputed criminal landscape, fuel trafficking would consolidate.

The Libyan Energy Context

After its civil conflict, by 2014 Libya was divided between warring factions from which two large groups standout: the Government of National

Accord (GNA) based in the west and the Libyan National Army (LNA) in the east. The GNA has the backing of the international community, while the LNA's control of the east gives it access to Libya's vast hydrocarbon reserves. Violent confrontations are common, 400,000 people have been internally displaced and approximately 11,000 had been killed by early 2020.

Libya has the largest oil reserves in Africa, yet the conflict undermined its energy sector. Oil barrel production decreased by 1.7 million units since the collapse of the Gaddafi regime, while the territorial division of the country between the LNA and the GNA has split Libya's national company (National Oil Corporation [NOC]) into two entities. The GNA-controlled western NOC is the only authorized to export oil and its derivates. In contrast, the LNA-controlled eastern NOC, continuously attempts to illegally obtain resources through unauthorized exports. As aforementioned, the eastern NOC has access to the majority of Libya's hydrocarbon resources (80%) and its potential profits represent an important motivation for the LNA to financially support its involvement in Libya's civil conflict.

The civil war not only undermined oil production it also reduced refining capacity. As a result, since 2011 Libya has had to import fuels. Between 2013 and 2017 the state spent $23.5 billion USD on imports (Observatory of Economic Complexity). To guarantee access to these derivates the state spent $30 billion USD on subsidies between 2012 and 2017 (ibid.). The subsidies and the security context have combined allowing for Libyan fuel traffickers to make substantial profits reselling derivatives in international markets.

Since 2014 the UN Security Council attempted to tackle the illegal exports of Libyan fuel with no results. That year, escalating conflict between the eastern and western factions led to pipelines, wells, terminals and refineries being repeatedly taken over by armed groups to exploit them (Bobin 2016). Approximately 30–40% of Libya's fuel, imported and nationally produced, is resold illegally in neighbouring and European countries annually. This represents losses for the Libyan state of approximately $750 million USD (Zaptia 2018).

Libyan authorities have shown great inconsistencies when it comes to measuring the scale of the country's fuels black market. Just like with the Mexican state, Libya shows no transparency of how exactly their figures are obtained and more rigorous scrutiny is needed to understand the scale of the problem. For example, in early 2017 the Attorney General stated that fuel trafficking losses amounted to $3.6 billion USD while never clarifying the period these losses corresponded to. Subsequently, official sources indicated that a whopping 85% of Libyan fuel supplies had been illegally diverted between January and November 2017. In contrast, the Libyan Audit Bureau placed the losses of fuel at $1.8 billion USD a year between 2013 and 2018 (Eaton 2018, 14). All these figures from official sources show a guesstimating trend by authorities about the fuel black market.

The Libyan Fuel Black Market

The Libyan Fuel Black Market (LFBM) operates on multiple levels: first, there is a branch of this illicit enterprise focused on transborder trafficking small quantities of fuel to neighbouring countries. Second, a national black market that diverts fuel within Libya to sell it at black-market rates and third, a transnational maritime-based trafficking involving industrial fuel quantities (Eaton 2018, 14). Border trafficking largely takes part with Tunisia, given the profit margins that the Libyan subsidies facilitate. In contrast, trafficking with neighbouring Egypt is less profitable because of subsidies implemented in that country. In Tunisia vehicle petrol costs $0.75 USD/litre while in Libya it costs $0.11 USD/litre (ibid.). Small-scale traffickers rely on illegality camouflaging using modified vehicles (oversized tanks, hidden compartments) to cross the Libyan border and make a profit in Tunisia.

After the collapse of the Gaddafi regime in 2011, fuel trafficking operations grew, as larger amounts of fuels started being stolen from tanker trucks, warehouses, ports and refineries. Exploiting strategic energy installations gives rise to larger fuel black markets, as the Mexican case will demonstrate. Large-scale fuel trafficking operations in Libya appear to be linked to the country's thriving war economy and relies on larger illicit networks to thrive. Illegally camouflaging using fake paperwork within this branch of the LFBM is widespread.

Within Libya, a network of informal service gas stations has been detected. These installations are the backbone for the LFBM national distribution network. In 2017 a spot-check by the NOC found that 87 out of 105 stations that were receiving fuel were not operating (Eaton 2018, 15). This inspection would eventually lead to signs of high-level co-option. In early 2018 NOC's chairman requested the Economy Ministry to end deliveries to ghost service stations, which was met with a dismissal. This response led the NOC's chairman to declare that this negative response reflected state capture and that fuel trafficking connections reached as high as the prime minister's office (ibid.).

Another cross-border overland fuel trafficking branch that emerged after 2011 involves the large-scale theft of tanker trucks. Most registered cases of stolen tanker trucks have taken place in southern Libya, an area where informal gas stations have outnumbered their legal counterparts. According to the UN Panel of Experts on Sudan, armed groups have been profiting by commercializing stolen Libyan fuel trucks in Darfur (Eaton 2018, 15). Lastly, we have the maritime trafficking-based side of the LFBM. These operations transcend the cross-border scope and have a transregional scale. In this context a vast fuel trafficking operation would consolidate, with a network that operated from Libya to the EU.

The Libyan-Maltese Connection

Malta is a regional strategic location for fuel transport. Situated between energy-consuming southern Europe and the energy-rich nations of Libya (oil) and Algeria (gas), its location at the crossroads of various maritime routes and the demand for its fuel storage terminals has made this island a strategic Mediterranean energy transport hub. Malta not only has a relevant location, it's also a place that over the past 15 years has consolidated as a criminal hub key to the functioning of different black markets (Raineri 2019, 11).

The reasons behind this consolidation are many and include the fact that since independence in 1964, Malta didn´t develop an efficient law enforcement and security apparatus. Malta also has an ambiguous jurisdictional status within the EU, maintaining a unique "regime of bank opacity" (Raineri 2019, 13) that provides individuals and companies the possibility of hiding their identities behind registered trusts. This situation has made Malta a gateway for illicit flows into Europe. Opaque transnational enterprises can set up operations in Malta using high-profile Maltese grey actors as fiduciaries. These agents, in turn, ease law enforcement scrutiny and guarantee impunity.

These characteristics would become relevant as the Libyan civil conflict worsened after 2011. Conflict affected all aspects of Libya's society, including its energy sector. In an earlier phase in 2012, the LFBM operated between neighbouring countries, as trafficking networks consolidated between Libya and Tunisia (Raineri 2019, 19). As aforementioned, criminals were exploiting the heavily subsidized Libyan fuel to obtain profits in Tunisia. Since its inception, the critical locations for these networks to operate had the potential to expand the black market beyond land borders. For example, the westernmost town of Zuwara had a port to sustain the operations of fuel tankers and Zawiya, to the east, had a refinery and a marine terminal. Libyan illicit groups realized that fuel trafficking generated larger profits than other black markets and represented lower risks. For example, it is estimated that this illicit enterprise generated revenues 2 to 4 times greater than human trafficking (Raineri 2019, 20).

Stigma and the possibility of state persecution also represented another incentive. One example occurred in Zuwara, which became a hub for Mediterranean human trafficking. In August 2015, a shipwreck departing from its port led to the death of 650 migrants. Public indignation led to a large government crackdown and opportunist violent retaliation amongst armed groups. This in turn drove criminal groups to relocate their human trafficking operations elsewhere or diversify into fuel trafficking (Eaton 2018, 31). This shows how loss of local legitimacy can compromise criminal operations and how state repression and violent illicit competition can lead to diversification to more profitable and secure black markets. In this context Malta would become a relevant protagonist as a connection for transnational fuel trafficking beyond North Africa. And here is where the criminal network, we are analysing, and its actors come into play.

The Network: Actors and Functions

An alleged key actor of this fuel smuggling network is a man known as Fahmi Ben Khalifa, a Libyan national who owned the business Tiuboda Refining Company. Khalifa was accused of fuel trafficking by Italian authorities (Guardia Di Finanza Di Catania 2017), but this institution was not alone in its indictment. The United Nations Panel of Experts on Libya accused Ben Khalifa since 2016 of leading one of the most relevant Libyan fuel trafficking networks (Farrah 2021). He was subsequently sanctioned by the Office of Foreign Assets Control (OFAC) of the US Department of the Treasury in 2018 for *threatening the peace, security, or stability of Libya through the illicit production, refining, brokering, sale, purchase, or export of Libyan oil* ("Treasury Sanctions International Network Smuggling Oil" 2018).

Being a grey actor, Ben Khalifa had presence in both the legal and the illicit realms: he was convicted for drug trafficking and was also investigated for involvement in human trafficking ("Italy Busts Gang" 2017). Ben Khalifa is a central network actor because of the resources and information he concentrated. This agent, whose actions we will analyse, exemplifies a few analytical tools of this research's theoretical framework: the participation of grey actors in fuel trafficking and black-market diversification (involved in "legitimate" energy trading, illegal hydrocarbons, narcotics and human trafficking).

Ben Khalifa had the support of a powerful grey actor who is part of Libya's volatile governing coalition. The support of this individual became evident in 2015, when Ben Khalifa was denounced for his involvement in fuel trafficking by a Libya-specialized journalist (Marlowe 2015). Back then Ali Faraj al-Qatrani (a politician linked to the LNA's leadership, member of a unity government with the GNA as a representative of the LNA in Libya's Presidential Council and head of the House of Representatives' Economy Committee) signed a letter affirming that Ben Khalifa's company, Tiuboda Refining Company, was legally registered and justified its operations as "oil services" (Zapita 2016; Eaton 2018, 16). The support by this high-ranking official allowed Ben Khalifa's trafficking operations to continue at the time, revealing the levels of state co-option that this network achieved and showing how a relevant grey actor can guarantee impunity and continuity for large-scale illicit operations.

Another relevant actor based in Libya is Tareq Dardar. According to Italian authorities, Dardar had experience in finance and is presumed to have provided money-laundering for Ben Khalifa's fuel trafficking through foreign accounts ("Italy Busts Gang" 2017; Guardia Di Finanza Di Catania 2017). Italian authorities stated that Dardar facilitated the flow of large sums of money from Tunisia to Libya, and to other countries in the Middle East and North Africa. He also supplied large amounts of cash to diverse agents of the network, especially to an armed group, the Shuhada al Nasr Brigade (Cimmarusti 2017). Dardar was also known for his alleged financial links to armed groups in Libya, showing that his participation in this fuel trafficking operation was one of many illicit activities he was involved in.

According to Italian investigators, Ben Khalifa collaborated with an armed group in Libya to access stolen fuel: the Shuhada al Nasr Brigade that controls the Zawiya Refinery 45 kilometres west of Tripoli. This installation is the second most important in Libya by refining capacity

(U.S. Energy Information Administration (EIA): Libya Overview 2020) and includes a port and an export terminal. Energy installations are key in fuel black markets, an issue we shall explore further. The Shuhada al Nasr Brigade collaborated with Ben Khalifa to divert imported gasoil that arrived at the Zawiya Refinery from Europe, using trucks to transport the fuel to the refinery's port terminal and to a makeshift fuel facility in Abu Kammash, 107 km to the west. In those locations, small fishing vessels took the gasoil to later set sail to deliver them to the next relevant actors in this network. According to the UN Panel of experts on Libya, fuel trafficking was one of the main revenue sources for armed groups like the Shuhada al Nasr Brigade in 2016 (United Nations Security Council 2016).

Before delving into the fuel trafficking network, it is important to further explain the Shuhada al Nasr Brigade. In Zawiya recent reports point that the group has 70 marine vessels and 30 illegal storage facilities for fuel smuggling (Raineri 2019, 20). The Brigade profits from multiple illegal enterprises, showing that the applicability of the black-market diversification component of this investigation has relevance beyond Mexico. According to the UN Security Council, the Shuhada al Nasr Brigade is the "*most dominant in the field of migrant smuggling and the exploitation of migrants in Libya*" ("Mohammed Kachlaf" 2018). The al Nasr Brigade is known for operating detention centres where it profits from migrant kidnapping and trafficking people for sexual exploitation. This armed faction has also participated in weapons trafficking.

The Shuhada al Nasr Brigade employs the support of grey actors. First there is the case of Ben Khalifa, an individual with presence in the formal energy sector and other illegal black markets that was key in enabling the Brigade to get the gasoil out of Libya. To get the fuel trafficking vessels into international waters and enable the capture of migrant boats, the leadership of this militia had "*extensive links with the head of the local unit of the coast guard of Zawiya*" ("Mohammed Kachlaf" 2018). This co-option between the Libyan coast guard and this group has been linked with the transnational growth of trafficked hydrocarbons from Libya since 2015 (Raineri 2019, 20). The support of these grey actors (in energy trade and state security) enabled this armed group to profit from diverse criminal markets.

The Shuhada al Nasr Brigade has shown great adaptability and a mercurial nature when it comes to its allegiances within Libya. The armed group has collaborated with actors from both the GNA and the LNA.

For example, the group is sanctioned by the GNA to "*provide security*" to the Zawiya Refinery ("Libya, January 2020 Monthly Forecast" 2020), while their criminal associate Ben Khalifa maintained links with a high-level military and political figure within the LNA (Vella 2017b). What this shows is the level of operational autonomy of the Shuhada al Nasr Brigade, allowing it to diversify into different black markets and sustain multiple alliances with different actors in a volatile environment.

Regarding fragmentation in 2012 the al Nasr Brigade was given control over the Zawiya Refinery, replacing the armed security wing of the NOC, the Petroleum Facility Guards (PFG). Eventually the al Nasr Brigade absorbed the PFG making both groups "*virtually indistinguishable*" (Raineri 2019, 19). Additionally, the chairman of the NOC admitted that his office had no control over PFG forces who operated autonomously (Eaton 2018, 16). What this assimilation points to is that Libyan armed groups are exposed to fragmentation in their volatile environment. Let us return to the fuel trafficking operation.

The vessels departed from the Zawiya Refinery sailing into the Libyan coast. Once they reached their destination, they would meet oil tankers from Malta. The ships linked to Ben Khalifa carried out these operations since at least 2014, switching off their transponder systems to erase any trace of their activities (Raineri 2019, 20). Intercepted vessels show the scale of these operations: in August 2017 two ships were detained off the Libyan coast transporting 6 million and 1.2 million litres of fuel respectively (Eaton 2018, 17). According to Italian authorities and the United Nations Panel of Experts on Libya, the vessels receiving the fuel off the Libyan coast belonged to two Maltese entrepreneurs: Darren Debono and Gordon Debono. Both businessmen (unrelated to each other) owned tankers and ships employed to transport stolen gasoil from Libya to Malta.

The United Nations Panel of Experts on Libya argued that the ships involved in these maritime trafficking operations sailed from the south of Malta to Libya and turned off their tracking systems to erase any traces. Once they loaded illicit derivatives, they returned to Malta. Before entering Maltese waters, the ships that arrived from Libya transferred to other vessels that took the shipments into the coast. The UN Panel mentioned two vessels involved that belonged to Ben Khalifa, accusing both ships of being detected on ten occasions alongside each other "*on the limits of Maltese territorial waters during 2015*" and detecting trips to Libya where both vessels shot down their detection systems "*following the pattern of smugglers*" (Panel of Experts on Libya 2016, 181). A ship owned by Darren Debono alone, under another company, was reportedly approached by two vessels owned by Khalifa on September 24, 2015, indicating fuel bunkering (maritime transfers between vessels)[3] (Vella 2017c; Panel of Experts on Libya 2016, 185).

Darren Debono was involved in the fishing industry and owned a seafood restaurant in the Maltese capital of La Valetta. Amongst the many business ventures owned by Darren Debono two stand out: Oceano Blu Trading Ltd and ADJ Trading Ltd. Oceano Blu was classified by investigators as one of the key companies for the criminal network to conduct its fuel smuggling operations (Guardia Di Finanza Di Catania 2017). In ADJ Trading, Darren Debono was an associate with Ben Khalifa. Company records show that Khalifa and Debono represented two-thirds of the corporation's ownership (Vella 2017a). This company owned two oil tankers that sailed to Libya regularly. According to Italian authorities Darren Debono was the actor that was "*dealing with the Libyan side of the business*" (ibid., 9).

Another relevant actor linked to Oceano Blu Trading is its former administrator, Italian national Nicola Orazio Romeo. Orazio Romeo took over the administration of this company since mid-2014 and, according to the Italian investigation, had connections with Mafia groups. This actor used Oceano Blu and other companies located in the British Virgin Islands to camouflage payments made to Ben Khalifa (Vella 2019). Darren Debono and Orazio Romeo knew each other before getting involved in fuel trafficking, when Debono's business ventures focused on fishing. Back then, according to investigative journalistic sources, Orazio Romeo helped his Maltese partner access a fish market controlled by Mafia groups in Catania to commercialize seafood (Rubino et al. 2018).

Italian prosecutors have accused Orazio Romeo of being an operative of the Santapaola-Ercolano family (Vella 2023), a criminal network with a strong presence in Catania (Rubino et al. 2018). Another Italian investigation included Romeo in its accused list for a money-laundering conspiracy that used a gaming company based in Malta with links to the same criminal group (Vella 2019, 2021, 2023). The fact that Romeo was also involved in a criminal gambling scheme is not fortuitous: for several years Malta has been known for its lax gambling regulations (allowing

---

[3] The UN Panel report presents as evidence the coordinates of the vessels involved in this encounter as well as other ships (Panel of Experts on Libya 2016,181).

the use of fiduciaries to circumvent controls) that have attracted criminal groups, particularly from Italy, to access cash and launder their illicit revenues, especially through online gambling (Raineri 2019, 17).

Finally, there is Gordon Debono, a fuel trader whose company Petroplus Ltd provided two tankers that transported fuel into Malta (Guardia Di Finanza Di Catania 2017). Gordon Debono provided the network with the knowledge and resources he gathered in the energy sector. According to investigations, he enabled acquiring certificates of origin from Saudi Arabia, used for illegality camouflaging and instructed on ship-to-ship maritime bunkering (Rubino et al. 2018). These three actors, the Debonos and Romeo, all had contacts in Sicily. The vessels owned by the Maltese operators oversaw bunkering the gasoil in Libya to take it into Malta or Sicily. In some cases, these ships would transfer their cargo to larger tankers (ibid., 8).

Once in Malta, the Libyan gasoil's formula was altered camouflaging its origin. According to Italian authorities this practice was effective in removing all traceability and enabled its resale. This was not the only measure taken to camouflage the illegality of these operations. Forged documentation was used in different moments of the fuel smuggling chain to simulate legality. This possibility of protecting illegal fuel trafficking operations, taking advantage of the fact that the smuggled goods are not inherently illegal, is not just present in this case involving the Mediterranean and continental Europe, it is also present in the Mexican context.

Documents show the level of sophistication this network had in camouflaging its fuel trafficking operations. This criminal network obtained certificates of origin from the Azzawiya Oil Refining Company (of the Zawiya Refinery) justifying the origin 50,000 metric tonnes of gasoil; from the Libyan–Maltese Chambers of Commerce stating that a shipment of Libyan gasoil came from Saudi Arabia; and from the business owned by Ben Khalifa, Tiuboda Oil Refining Company (Illustration 2.1).

Having explained the varied roles and methods behind this criminal network (actors occupying positions in legal and illegal realms, illegal market diversification, etc.), we are still missing the role played by Maltese authorities. Public sector agents can be crucial grey actors given the strategic resources they provide to criminal networks (impunity, access to installations, documentation, classified information, distorting institutional processes, accessing public budgets and so on). In Malta, though no formal accusations have been made against state actors, many

Illustration 2.1 Certificates of origin used by the Ben Khalifa/Debono Network (Guardia Di Finanza Di Catania 2017)

sources point to the authorities' inaction (and in some cases participation) involving the operations of this fuel trafficking network. Maltese officials can't argue this situation caught them by surprise: since 2016 Libyan and UN authorities were warning about the involvement of the Debonos and Ben Khalifa in fuel trafficking (Raineri 2019, 21; Panel of Experts on Libya 2016, 179–186).

Fishing cooperatives complained that the ships owned by Darren and Gordon Debono performed their bunkering activities in Maltese waters uninterrupted and could dock without inspections by custom authorities ("Multi-million Fuel Smuggling Operation" 2018). An employee of the Ministry of Foreign Affairs in Malta certified the authenticity of the certificates used by this network to camouflage the origins of their illegally obtained gasoil (ibid.). Additionally, we must consider the Maltese context in which criminal and political collusion is commonplace. In 2016 for example, the son of a former prime minister was amongst the shareholders of a trust owning a betting society controlled by the Sicilian Mafia

(Raineri 2019, 17). Darren Debono had a close relationship with Malta's prime minister Joseph Muscat (ibid., 21).

So far, we have explained the agents that made up the Libyan–Maltese nodes of this network, but they were not the only ones. There are also those who, allegedly, facilitated access to continental Europe.

The Continental Actors

This criminal network reached a cross-regional scope, operating in North Africa and Europe. By doing this, this group of individuals got past different cultural backgrounds, language barriers and various national jurisdictions to obtain illicit profits. And according to Italian authorities, an energy company paid an instrumental role in extending the reach of this network into the European continent: Maxcom Bunker SA.

Maxcom, at the time of these illicit operations,[4] was a company dedicated to international wholesale distribution of petroleum products. Italian investigators argue that it received four ships with 82,000 tonnes of trafficked fuel belonging to the Debonos between June 2015 and June 2016. Maxcom Bunker allegedly acquired these shipments for €27 million, when market value reached €51 million (Guardia Di Finanza Di Catania 2017). It is believed that this Italian fuel trading company acquired the gasoil for €0.28/litre while the legal price was €0.53/litre.

This company is also accused of adulterating the Libyan gasoil's formula. According to authorities, this allowed the company to bypass EU regulations (which restrict this derivative to maritime use) and to commercialize it in Italian, Spanish and French gas stations through camouflaged distribution (ibid.). If these accusations are confirmed, and by selling their illegal fuel in legitimate stations, this network deployed an "*invisible supply chain*", exploiting the legal market (Ralby et al. 2019). Eventually, Italian authorities would detain the company's Managing Director, Marco Porta, for his alleged involvement in the Ben Khalifa/ Debono network ("Italian Police Bust Libyan" 2017).

---

[4] According to a bunkering specialized publication Maxcom Bunker was formally replaced by Bunker Energy in late 2018 (https://issuu.com/constructivemedia/docs/wb_spring_issue_2019).

Furthermore, According to the UK Government, Maxcom Bunker is not operating as a company (https://find-and-update.company-information.service.gov.uk/officers/sx2e5W3wEGnp_ZEJm_HseLHkwpc/appointments).

This investigation revealed valuable information of the roles and resources supposedly brought by Maxcom for this criminal operation. Besides adulterating the Libyan gasoil's formula, this company is also accused of camouflaging their operations through the use of falsified certificates of origin, bills of lading and deploying corporate and financial coverups (Guardia Di Finanza Di Catania 2017). Within Maxcom Bunker, Marco Porta is suspected of managing this illicit operation. A business office employee and Italian national purportedly acted as a link with the Maltese and Sicilian actors of the network. An Italian consultant stands accused of organizing meetings and developing relationships within the network; while a company employee allegedly managed a warehouse in Augusta, Sicily where the adulterated gasoil was brought to triangulate distribution in Sicily and in mainland Italy (ibid.). This information exemplifies some functions that individuals can play as private sector actors within secretive criminal networks.

This alleged involvement of Maxcom Bunker can exemplify the importance that legal companies can have as grey actors. Legally constituted companies are capable of providing fuel trafficking networks with transport, storage, illegality camouflaging and the capacity to exploit legal supply chains. Access to legitimate markets can in turn enable generating vast profits through camouflaged commercialization. This capacity to expand the scope and revenues of criminal operations can make legal enterprises network-extending grey actors (Williams 2001, 81), which are also present in the Mexican case. Additionally, through the use of invisible supply chains "legal" actors and entities could unknowingly be participating in fuel smuggling operations, fulfilling a different array of functions for illicit networks.

The following chart shows the key actors of the Ben Khalifa/Debono network according to investigations, showing its complex structure and its sophisticated operations (Illustration 2.2).

The Lessons of the Ben Khalifa/Debono Network

Between 2015 and 2016 accusations against Ben Khalifa were made, including one made by the UN. Yet the network's powerful grey actors gave it impunity to keep operating until late 2017. An example of this is when Ali Faraj al-Qatrani, the representative of the LNA in Libya's Presidential Council, vindicated Ben Khalifa's fuel trafficking operations in 2015. Subsequently, al-Qatrani retracted his support for Ben Khalifa

2 AN INTRODUCTION TO FUEL TRAFFICKING 47

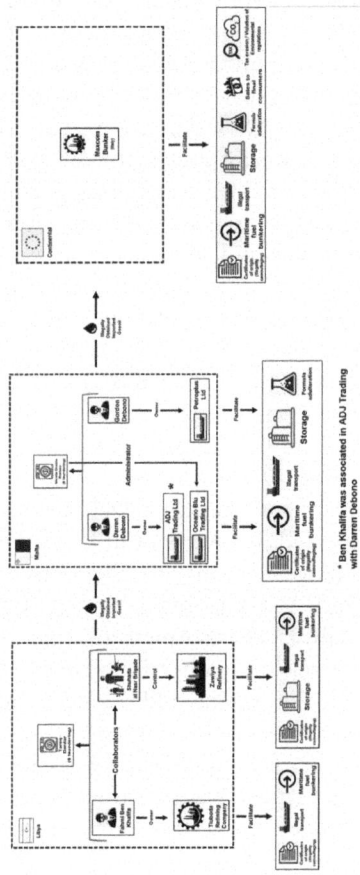

Illustration 2.2 Actors and functions of the Ben Khalifa/Debono Network (Elaborated by author)

stating (vaguely) that private companies could not export Libyan derivatives (Zapita 2016). The withdrawal of this powerful actor's support allowed for the detention of Ben Khalifa in Libya. In August 2017 it was reported that Ben Khalifa was detained by Tripoli's Rada Special Deterrence Force, a security institution under the Libyan Ministry of the Interior (Rubino et al. 2018). The group declared that they carried out the detention due to Khalifa's alleged involvement in fuel and human trafficking. Journalistic sources framed this detention within a renewed European effort to strengthen Libya's capacity to patrol its territorial waters (Vella 2017c). What Khalifa's detention after losing high-level support exemplifies is how a powerful grey actor can guarantee impunity for a criminal network and its core nodes enabling its continuity.

A similar case occurred in Malta, where authorities were slow to tackle the Ben Khalifa/Debono network and where state actors (custom authorities) showed direct involvement in their fuel trafficking operations by accepting and validating documentation. Additionally, Malta's role as a Mediterranean criminal hub cannot be ignored. An example that reinforces this argument is the fact that the detention of key actors of the Ben Khalifa/Debono network (consequence of an Italian investigation) had no impact on other fuel trafficking groups in Malta well into 2018 (Raineri 2019, 21).

In October 2017 Darren and Gordon Debono were arrested in Catania, Italy, for their suspected involvement in fuel smuggling (Montebello 2023). In 2018, Ben Khalifa and the Debonos were sanctioned by the US Office of Foreign Assets Control for the connection to fuel smuggling originating in Libya. These American sanctions included 24 companies, seven marine vessels and other individuals ("Treasury Sanctions International Network Smuggling Oil" 2018). In Malta, by late 2020, Gordon and Darren Debono were charged with money-laundering apparently derived from fuel trafficking (Montebello 2023).

This case shows the roles that established companies can play as parts of fuel trafficking networks, enabling transport, technical know-how, storage, adulteration, camouflaged documentation, money-laundering, transnational distribution, legal supply chain capture and so on. More generally, private companies involved in mining, extraction and other natural resources can be crucial grey agents for criminal networks. Research on the criminal exploitation of timber has pointed out the importance of legal companies in this commodity chain "*giving a legal appearance to the illegally harvested timber*" and enabling large profits through exports (Boekhout van Solinge et al. 2016, 86). Therefore, these companies should be considered when analysing extractive illegal markets.

The point of this case study is to stress that this investigation's theoretical framework and analytical tools can be useful for analysing international contexts. By briefly listing the analytical tools of this investigation and applying them to this case we found:

I. The participation of diverse criminal networks: Different actors belonging to various criminal groups all took part in this case: Libyan armed groups, Libyan and Maltese businessmen, money-laundering actors, politicians, bureaucrats, Mafia groups and legal enterprises based in continental Europe. This example shows the participation of different criminal networks and agents profiting from fuel trafficking beyond Mexico.
II. The presence of grey actors: All over this case we see repeatedly the participation of grey actors that enable and amplify the scope of this fuel trafficking network. Obtaining documentation, enabling transport, storage, adulteration, distribution, guaranteeing impunity; without these grey-actor resources the Ben Khalifa/Debono network could not have operated for so long and reached the scope it accomplished.
III. Illegality camouflaging: Using falsified certificates of origin, façade companies and the adulteration of original formulas, the case of the Ben Khalifa/Debono network depicts the crucial protection that illegality camouflaging plays in fuel trafficking.
IV. Black-market diversification: Most criminal agents involved in this case are accused of participating in diverse black markets: Ben Khalifa (drug and human trafficking), Shuhada al Nasr Brigade (human and migrant trafficking, weapons smuggling), Tareq Dardar (money-laundering, financial support of armed groups) and Orazio Romeo (money-laundering). All these examples show that the analytical tool of black-market diversification is useful in diverse contexts.
V. Fragmentation of criminal groups: Just like in Mexico, Libya had a regime that regulated a thriving criminal underworld. The collapse of the Libyan regime disrupted the mechanisms that structured the criminal realm. The resulting scenario led to large-scale violent competition that created a context where criminal fragmentation became cyclical.

VI. Criminal co-option: the case of the Ben Khalifa/Debono network shows the far-reaching implications that criminal co-option of state and non-state actors can have in sophisticated fuel trafficking. Officials, bureaucrats, entrepreneurs, armed groups, private companies and Mafia members; all these actors coordinated implementing illegal actions and sustained agreements that allowed them to obtain continuous benefits evading law enforcement across different jurisdictions. This is another analytical tool that transcends Mexico.

The case of the Ben Khalifa/Debono network shows the comparative value of this research's theoretical framework. By developing the case of the MFBM in the chapters ahead, this investigation builds a set of analytical tools that can contribute to examine other instances of hydrocarbons trafficking elsewhere in the world. This is by no means a definitive framework and I believe that further research can contribute to refine and expand it.

Bibliography

"Bulgaria's Transport Minister Orders Inquiry into Suspected Diesel Theft at State Railways." Novinite, January 26, 2015. https://www.novinite.com/articles/166170/Bulgaria%E2%80%99s+Transport+Minister+Orders+Inquiry+into+Suspected+Diesel+Theft+at+State+Railways. Date accessed: July 11, 2021.

"#Catania. #Riciclaggio di gasolio illecitamente prelevato e destinato ad essere immesso in consumo come carburante: 9 soggetti #arrestati." 2017. Video. https://twitter.com/GDF/status/920583374623924224?ref_src=twsrc%5Etfw%7Ctwcamp%5Etweetembed%7Ctwterm%5E920583374623924224&ref_url=https%3A%2F%2Ffr.euronews.com%2F2017%2F10%2F19%2Fhalte-au-trafic-de-gasoil-frelate-en-europe. Date accessed: July 12, 2021.

"Fuel Fraud Costing Europe More Than $4 Billion in Lost Taxes." Bloomberg. August 27, 2013.

"Italian Police Bust Libyan Fuel-Smuggling Ring." The Maritime Executive, October 18, 2017. https://www.maritime-executive.com/article/italian-investigators-bust-libyan-fuel-smuggling-ring. Date accessed: July 11, 2021.

"Italy Busts Gang Smuggling Zawia Diesel to Sicily." *Libya Herald*, October 18, 2017. https://www.libyaherald.com/2017/10/18/italy-busts-gang-smuggling-zawia-diesel-to-sicily/. Date accessed: July 12, 2021.

"Libya, January 2020 Monthly Forecast." 2020. Security Council Report. https://www.securitycouncilreport.org/monthly-forecast/2020-01/libya-10.php. Date accessed: July 12, 2021.
"Maxcom Bunker SpA." Bloomberg. https://www.bloomberg.com/profile/company/4337893Z:IM. Date accessed: July 12, 2021.
"Mohammed Kachlaf." 2018. United Nations Security Council. https://www.un.org/securitycouncil/sanctions/1970/materials/summaries/individual/mohammed-kachlaf. Date accessed: July 12, 2021.
"Multi-Million Fuel Smuggling Operation Coordinated from Malta." *Times of Malta*, May 3, 2018. https://www.timesofmalta.com/articles/view/20180503/local/multi-million-fuel-smuggling-operation-coordinated-from-malta.678130. Date accessed: July 11, 2021.
"One More Month in Jail for Ex-Investigation Chief." *Bangkok Post*, February 27, 2018. https://www.bangkokpost.com/thailand/general/1419274/one-more-month-in-jail-for-ex-investigation-chief. Date accessed: July 11, 2021.
"Police Chief Says Oil Smugglers Funding Insurgency." Thai PBS, September 13, 2013.
"Treasury Sanctions International Network Smuggling Oil from Libya to Europe." US Department of the Treasury, February 26, 2018. https://home.treasury.gov/news/press-releases/sm0298. Date accessed: April 15, 2024.
"Venezuela: A Mafia State?" Insight Crime, May 2018: 1–84.
Aretxaga, Begoña. 2003. "Maddening States." *Annual Review of Anthropology* 32: 393–410.
Atuesta, Laura, and Yocelyn Pérez-Dávila. 2018. "Fragmentation and Cooperation: The Evolution of Organized Crime in Mexico." *Trends in Organized Crime* 21: 235–261.
Beckert, Jens, and Matías Dewey. 2017. "The Social Organization of Illegal Markets." In *The Architecture of Illegal Markets: Towards an Economic Sociology of Illegality in the Economy*, edited by Jens Beckert and Matías Dewey, 1–38. Oxford University Press.
Beckert, Jens, and Frank Wehinger. 2013. "In the Shadow: ILLEGAL MARKets and economic Sociology." *Socio-Economic Review* 11 (1) (January): 5–30.
Bobin, Frédéric. 2016. "En Libye, l'incessante bataille du pétrole." Le Monde Afrique, September 26. Date accessed: December 19, 2023.
Boekhout van Solinge, Tim, Pieter Zuidema, Mart Vlam, Omar Cerutti Paolo, and Valentin Yemelin. 2016. "Organized Forest Crime: A Criminological Analysis with Suggestions from Timber Forensics." *International Union of Forest Research Organizations*: 81–96.
Cimmarusti, Ivan. 2017. "Petrolio illecito dalla Libia, business da 51 milioni." *Il Sole 24 Ore*, October 19. https://www.ilsole24ore.com/art/petrolio-illecito-libia-business-51-milioni-AEjijOrC. Date accessed: July 12, 2021.

Durán-Martínez, Angélica. 2018. *The Politics of Drug Violence: Criminals, Cops and Politicians in Colombia and Mexico*, 1–299. Oxford University Press.

Eaton Tim. 2018. *"Libya's War Economy Predation, Profiteering and State Weakness*, 2–34." Chatham House, April.

Enciso, Froylán. 2017. "México y la guerra sin nombre." Crisis Group, June 15. https://www.crisisgroup.org/es/latin-america-caribbean/mexico/mexicos-worsening-war-without-name. Date accessed: July 11, 2021.

Farrah, Raouf. 2021. "Zuwara's Civil Society Fight Against Organized Crime: Successes and Failures of Local Community Efforts." Global Initiative Against Transnational Organized Crime, December 2021. https://globalinitiative.net/wp-content/uploads/2021/12/Zuwara-PB-web.pdf. Date accessed: April 15, 2024.

Felbab-Brown, Vanda. 2019a. *Mexico's Out-Of-Control Criminal Market*, 1–29. The Brookings Institution.

Fuerte Celis, María del Pilar, Enrique Pérez Lujan, and Rodrigo Cordova Ponce. 2018. "Organized Crime, Violence, and Territorial Dispute in Mexico (2007–2011)." *Trends in Organized Crime* 22: 188–209.

Garay-Salamanca, Luis Jorge, and Eduardo Salcedo-Albarán. 2016. *Macro-Criminalidad: Complejidad y Resiliencia de las Redes Criminales*, 1–191. iUniverse.

Garay-Salamanca, Luis, Eduardo Salcedo-Albarán, Guillermo Macías Fernández, Diana Santos Cubides, and Nathalia Guerra Villamizar. 2018. *Macro-Corruption and Institutional Co-Optation: The "Lava Jato" Criminal Network*, 13–209. Vortex Foundation.

Garzón, Juan Carlos. 2014. "From Drug Cartels to Predatory Micro Networks: The 'New' Face of Organized Crime in Latin America." In *Reconceptualizing Security in the Western Hemisphere in the 21st Century*, edited by Bruce M. Bagley, Jonathan D. Rosen, and Hanna Kassab, 117–131.

Gingeras, Ryan. 2014. "Corruption, Crime, and Scandal in Turkey." https://blog.oup.com/2014/11/corruption-smuggling-turkey-government/. Date accessed: July 11, 2021.

Guardia Di Finanza Di Catania. 2017. "Operazione "Dirty Oil"." October 18, 2017. https://www.publiceye.ch/fileadmin/doc/Rohstoffe/Guardia-di-Finanza-di-Catania_Dirty-Oil_La-complessa-investigazione-economico.pdf.Date accessed: July 12, 2021.

Gustafsson, Johanna. 2017. *Single Case Studies vs. Multiple Case Studies: A Comparative Study*, 1–15. Academy of Business, Engineering and Science, Halmstad University.

Happy, Boris Odalonu. 2015. "The Upsurge of Oil Theft and Illegal Bunkering in the Niger Delta Region of Nigeria: Is There a Way Out?" *Mediterranean Journal of Social Sciences*, 2nd ser., 6 (02) (May): 563–573.

Hernández, Enrique. 2019. "Ducto Tuxpan-Azcapotzalco, la mina de oro negro de los cárteles." El Sol de México, January 16. https://www.elsoldemexico.com.mx/mexico/justicia/desabasto-de-gasolina-por-que-perforan-ducto-tuxpan-azcapotzalco-mina-de-oro-negro-carteles-huachicoleo-pemex-2929001.html. Date accessed: July 11, 2021.

Hübschle, Annette. 2017. "Contested Illegality: Processing the Trade Prohibition of Rhino Horn." In *The Architecture of Illegal Markets: Towards an Economic Sociology of Illegality in the Economy*, edited by Jens Beckert and Matías Dewey, 1–22. Oxford University Press.

Ikelegbe, Augustine. 2005. "The Political Economy of Conflict in the Oil Rich Niger Delta Region of Nigeria." *Nordic Journal of African Studies* 14 (2): 208–234.

Imagen Noticias. 2019. "Acompañamos a los marinos que recorren a pie los caminos del huachicol." Video. https://www.youtube.com/watch?v=ZB92b_Y3jZ0&t=114s. Date accessed: July 11, 2021.

Judith Burdin, Asuni. "Blood Oil in the Niger Delta." United States Institute of Peace Special Report 229. August 2009, 1–16.

Kiourktsoglou, George, and Alec Coutroubis. 2015. "ISIS Export Gateway to Global Crude Oil Markets." Maritime Security Review, 1–16. http://www.marsecreview.com/wp-content/uploads/2015/03/PAPER-on-CRUDE-OIL-and-ISIS.pdf. Date accessed: July 11, 2021.

Lessing, Benjamin. 2015. "Logics of Violence in Criminal War." *Journal of Conflict Resolution* 59 (8) (December): 1486–1516.

Marlowe, Ann. 2015. "Why Does EU Tolerate Libya's Smuggler Kingpin as Migrants Drown?" *Asia Times*, October 16. https://asiatimes.com/2015/10/eu-turns-blind-eye-to-fuel-for-arms-smuggling-as-migrants-drown/. Date accessed: July 12, 2021.

Mayntz, Renate. 2017. "Illegal Markets Boundaries and Interfaces Between Legality and Illegality." In *The Architecture of Illegal Markets: Towards an Economic Sociology of Illegality in the Economy*, edited by Jens Beckert and Matías Dewey, 1–11. Oxford University Press.

McAleese, Deborah. 2016. "Launderers Turn to Fuel Smuggling After Diesel Marker Stumps Them." *Belfast Telegraph*, March 28. https://www.belfasttelegraph.co.uk/news/northern-ireland/launderers-turn-to-fuel-smuggling-after-diesel-marker-stumps-them-34576526.html. Date accessed: July 11, 2021.

Medel, Monica, and Francisco Thoumi. 2014. "Mexican Drug "Cartels"." In *The Oxford Handbook of Organized Crime*, edited by Letizia Paoli, \ 1–26. Oxford University Press.

Montebello, Sean. 2023. "Gordon Debono using luxury yacht despite asset freeze." The Shift, October 22. https://theshiftnews.com/2023/10/22/gordon-debono-using-luxury-yacht-despite-asset-freeze/. Date accessed: April 15, 2024.

Observatory of Economic Complexity. Libya Imports 2013–2017. https://oec. world/en/profile/country/lby#Imports. Date accessed: December 19, 2023.
Pachico, Elyssa. 2011. "Fuel-Theft in Latin America: A Fail-Safe Trade?" *Insight Crime*, May 10. https://www.insightcrime.org/investigations/fuel-theft-in-latin-america-a-fail-safe-trade/. Date accessed: July 11, 2021.
Panel of Experts on Libya. 2016. "Letter Dated 4 March 2016 from the Panel of Experts on Libya Established Pursuant to Resolution 1973 (2011) Addressed to the President of the Security Council." Report No. S/2016/209. United Nations Security Council, 2016. https://www.securitycouncilreport.org/atf/cf/%7B65BFCF9B-6D27-4E9C-8CD3-CF6E4FF96FF9%7D/s_2016_209.pdf.
Park, Jung H. 2016. "What Explains the Patterns of Diversification in Drug Trafficking Organizations?" Master's thesis, Naval Postgraduate School, Monterrey California, 1–85.
Pérez, Ana Lilia. 2011. *El Cártel Negro: Cómo el crimen organizado se ha apoderado de Pemex*, 5–221. México: Grijalbo.
Raineri. Luca. 2019. "The Malta Connection: A Corrupting Island in a "Corrupting Sea"?" *The European Review of Organised Crime* 5 (1): 10–35.
Ralby, Ian M. 2017. *Downstream Oil Theft: Global Modalities, Trends and Remedies*, 1–117. Atlantic Council Global Energy Centre, January.
Ralby, Ian, David Soud, and Rohini Ralby. 2019. "Defining the Invisible Supply Chain." Atlantic Council, March 20. https://www.atlanticcouncil. org/blogs/energysource/defining-the-invisible-supply-chain. Date accessed: July 11, 2021.
Rubino, Giulio, Cecilia Anesi, and Lorenzo Bagnoli. 2018. "Death in a Smugglers' Paradise." Organized Crime and Corruption Project, May 3. https://www.occrp.org/en/thedaphneproject/death-in-a-smugglers-paradise. Date accessed: July 12, 2021.
Rufyikiri, Gervais. 2016. "Grand Corruption in Burundi: A Collective Action Problem Which Poses Major Challenges for Governance Reforms." Institute of Development Policy and Management, University of Antwerp, April: 5–18.
Stephen, Chris. 2014. "War in Libya—The Guardian Briefing." *The Guardian*, August 29. https://www.theguardian.com/world/2014/aug/29/-sp-briefing-war-in-libya. Date accessed: July 12, 2021.
Sullivan, John, and Adam Elkus. 2011. "Open Veins of Mexico: The Strategic Logic of Cartel Resource Extraction and PetroTargeting." *Small Wars Journal*, November 3: 1–9.
Thepbamrung, Nattha. 2013. "Diesel Smugglers Arrested Off Phuket." Phuket News, February 8. https://www.thephuketnews.com/diesel-smugglers-arrested-off-phuket-36939.php#ilKoS1Rq2ZGt81cx.97. Date accessed: July 11, 2021.

United Nations Security Council. 2016. "Final Report of the Panel of Experts on Libya Established Pursuant to Resolution 1973 (2011)." March 9. https://www.securitycouncilreport.org/atf/cf/%7B65BFCF9B-6D27-4E9C-8CD3-CF6E4FF96FF9%7D/s_2016_209.pdf.
U.S. Energy Information Administration (EIA). 2020. Libya Overview. https://www.eia.gov/international/analysis/country/LBY. Date accessed: July 12, 2021.
Vella, Matthew. 2017a. "Maltese Wanted in Italy over Libyan Fuel Smuggling Racket Worth €30 Million." *Malta Today*, October 18. https://www.maltatoday.com.mt/news/national/81428/maltese_wanted_in_italy_over_libya_fuel_smuggling_racket. Date accessed: July 11, 2021.
Vella, Matthew. 2017b. "Smuggling Kingpin Got Libyan Politician's 'stamp of Approval' for Fuel Exports." Libya Tribune, November 24. https://en.minbarlibya.org/2017/11/24/smuggling-kingpin-got-libyan-politicians-stamp-of-approval-for-fuel-exports/. Date accessed: December 19, 2023.
Vella, Matthew. 2017c. "Kingpin Fuel and Human Smuggler with Malta Links Arrested by Libyans." *Malta Today*, August 28. https://www.maltatoday.com.mt/news/national/80035/kingpin_smuggler_with_malta_links_arrested_by_libyans. Date accessed: April 15, 2024.
Vella, Matthew. 2019. "Sicilian Oil Smuggling Suspect Is 'No Mafioso', Lawyer Warns." *Malta Today*, October 14. https://www.maltatoday.com.mt/news/national/97985/sicilian_oil_smuggling_suspect_is_no_mafioso_lawyer_warns_maltatoday#.Xe5O0JKhTY. Date accessed: July 12, 2021.
Vella, Matthew. 2021. "Italian Mafia Bust Reveals Deep Santapaola Gaming Links to Malta." *Malta Today*, March 15. https://www.maltatoday.com.mt/news/national/108335/italian_mafia_bust_reveals_deep_santapaola_gaming_links_to_malta. Date accessed: April 15, 2024.
Vella, Matthew. 2023. "Anti-Mafia Cops Tell Italian MPs of Malta Base for Fugitives." *Malta Today*, July 3. https://www.maltatoday.com.mt/news/national/123730/antimafia_cops_tell_italian_mps_of_malta_base_for_fugitives_. Date accessed: April 15, 2024.
Vigil, Michael. 2016. "The Structure and Psychology of Drug Cartels." The CIPHER Brief, June 15, 2016. https://www.thecipherbrief.com/column_article/the-structure-and-psychology-of-drug-cartels. Date accessed: July 11, 2021.
Villalba, Javier. 2018. "New Criminal Group Runs Fuel Smuggling at Colombia-Venezuela Border." Insight Crime, August 29, 2018. https://www.insightcrime.org/news/analysis/new-criminal-group-runs-fuel-smuggling-colombia-venezuela-border/. Date accessed: July 10, 2021.
Williams, Phil. 2001. "Transnational Criminal Networks." In *The Future of Terror, Crime, and Militancy*, edited by John Arquilla, David Ronfeldt, 61–97. RAND Corporation.

Willis, Goddey. 2014. "The Nigerian State and Oil Theft in The Niger Delta Region of Nigeria." *Journal of Sustainable Development in Africa* 16 (1): 69–81.

Yashar, Deborah J. 2018. *"Homicidal Ecologies: Illicit Economies and Complicit States in Latin America*, 1–368. Cambridge University Press.

Zapita, Sami. 2016. "Gatrani Alerts Maltese Government to the Illegality of Fuel Smuggling from Libya to Malta." *Libya Herald*, July 21. https://www.libyaherald.com/2016/07/21/gatrani-alerts-maltese-government-to-the-illegality-of-fuel-smuggling-from-libya-to-malta/. Date accessed: July 12, 2021.

Zaptia, Sami. 2018. "$750 m Worth of Libyan Fuel Is Stolen: Sanalla." *Libya Herald*, April 19. https://libyaherald.com/2018/04/750-m-worth-of-libyan-fuel-is-stolen-sanalla/. Date accessed: December 19, 2023.

PART II

La Sierra Norte de Puebla

Mist covers the landscape, giving it an ethereal quality. A military truck appears, its cargo area full of sitting marines. The low visibility doesn´t stop the convoy as it pierces through the fog with its sirens´ lights on. We are in *Xicotepec* ("Place of the Bumblebees" in Nahuatl) in *la Sierra Norte de Puebla*. It is an exuberantly lush site, an evergreen tropical forest extending as far as the eyes can see. Suddenly, the marines are proceeding by foot along unpaved roads that are barely holding their shape as the greenery erodes them.

The marines are here to patrol the *Tuxpan–Azcapotzalco* polyduct pipeline, a 311-kilometre installation that connects Veracruz to Mexico City (Hernández 2019) and that ascends more than a thousand meters to serpent through Xicotepec´s rugged mountain region. The marines look for the pipeline among the lush vegetation. The serpentine pipeline appears in some sections at ground level while hiding underground in others. "This was reported yesterday" says a PEMEX worker escorted by the marines while pointing to an illegal extraction point that is barely noticeable in the dirt. "I think they will repair it Saturday; they give it less priority here because the problem is bigger in Tula" he remarks.

The Marines act as trackers to find illegal extraction points in this challenging place, relying on discovering vehicle tracks, picking up the scent of fuel or hearing pressure leaks. They rigorously patrol the areas where the *Tuxpan–Azcapotzalco* polyduct is exposed at surface level. In these points, discovered illegal extraction points are reactivated by traffickers less than six days after being sealed (Imagen Noticias 2019). Marines have detected up to ten extraction points in less than a day. Their findings reflect the resourcefulness of the criminals behind pipeline fuel-theft: from quick closing valves (that activate and deactivate the flow of fuels immediately) to hoses that supply vehicles at road level. The journalist accompanying the marines on their patrol asks, astonished, "So, they load fuel in the middle of the road?". "Yes, and in a flash" answers the distorted voice of the PEMEX worker.

References

Hernández Enrique. 2019. "Ducto Tuxpan-Azcapotzalco, la mina de oro negro de los cárteles". El Sol de México. January 16, 2019. https://www.elsoldemexico.com.mx/mexico/justicia/desabasto-de-gasolina-por-que-perforan-ducto-tuxpan-azcapotzalco-mina-de-oro-negro-carteles-huachicoleo-pemex-2929001.html . Date Accessed July 11, 2021.

Imagen Noticias. 2019. "Acompañamos a los marinos que recorren a pie los caminos del huachicol". Video. https://www.youtube.com/watch?v=ZB92b_Y3jZ0&t=114s . Date Accessed July 11, 2021.

CHAPTER 3

The Mexican Context

This chapter will explore the energy and security context in which Mexican fuel trafficking first appeared. Afterwards, the origins of this illicit activity and its beginning as a relevant criminal enterprise will be explored. The size of the MFBM, the modalities of theft and the criminal actors involved will subsequently be developed. This chapter will show that, though recent, fuel trafficking was noticed before its 2011–2018 expansion and that PEMEX had a deep-rooted corruption history that made it an ideal co-option candidate.

Mexico's Energy and Security: A Tale of Two Crises

The Energy Crisis

Since the discovery of the first oilfield at the start of the twentieth century oil became a vital component of Mexico's politics, economy and society. Its importance grew during the presidency of Lázaro Cárdenas (1934–1940), when the nationalization of the oil industry heralded a historic change for Mexico, altering the balance of power from the military to the parastatal sector. This enabled the state to consolidate itself as the managing entity of a mixed economy and marked the beginning of the rule of a corporative pact between unions and the one-party regime (González 2001). The national oil industry had historical significance beyond its conception, reinforced by a period from 1938 to the mid-1970s in which the sector became the pillar of Mexico's industrialization, enabling the appearance of an agricultural sector through the production of fertilizers, and permitting the production of cement, iron and steel (Rousseau 2010, 307). This industry shaped a large part of the one-party rule and became a cornerstone for Mexico's modernization. Its importance is earth-shaking.

Despite this importance, PEMEX (a monopoly since 1940) struggled with several challenges. The success story that made PEMEX a matter of national pride would deteriorate beyond recognition because of the prioritization of the company's fiscal and social roles over its industrial necessities. The importance given to those priorities would lead to restricted expenses on maintenance, imports, spare parts or non-productive investments in the short term (this led to insufficient exploration that later became a pressing concern). In 1977, national energy policy took a turn. While it was at first the basis of the industrialization and development of the country, oil became a guarantor for international credit for an exhausted political system. Two events allowed this to happen: the discovery of giant oilfields in Mexico's south-east and the record oil barrel prices in international markets. These events petrolized Mexico's economy. In 1982 this model collapsed with the sharp decrease in oil prices and soaring interest rates. The country was on the verge of bankruptcy, caused by the very sector that shaped its modernization decades earlier.

Besides these contradictions, PEMEX has faced since its inception a growing problem of internal corruption, led by the powerful oil workers union (STPRM). The union leadership had historically occupied positions of political representation, enabling their access and management of oil resources to support corporatist structures. Access to these resources was sustained through corrupt agreements with PEMEX's political leadership (González 2001, 122). When the nationalization of the industry was underway, union leaders took over managerial positions across Mexico's energy installations without renouncing to their labour control. This conflicting dichotomy in the commanding heights of PEMEX created a linkage between "*the interests of the company, the union interests and the particular interests of the officials, generating in a short time a structure of corruption that influences the functioning of the industry and determines the relations between the union and PEMEX*" (ibid., 125–126).

Since the formalization of the collective labour agreement in 1947, the union leadership has been granted discretionary concessions like controlling the hiring of workers. This allowed the labour leaders, involved with officials, to sell job positions. Another example of corruption was clause 36 of the same agreement. This clause allowed PEMEX to outsource projects, such as installation construction, transport, exploration and drilling. STPRM's leaders profited from this clause by becoming middlemen or through direct involvement via contracted companies they owned (ibid.). These concessions accumulated, allowing the union leadership and officials to enrich themselves unlawfully becoming contractors or obtaining bribes to assign contracts. The vast oil resources, though the majority remained at the top, permeated all the way through the union's corporatist structures including lower-level workers. The system was plentiful but fragile, depending on the capacity of the industry to remain profitable.

Subsequent reform attempts tried to tackle the contradictions within PEMEX, divided between its purposes of being a political and fiscal cash cow and an oil company needing to tend to its investments and profitability. These reforms far from correcting the problems, worsened them, or created new ones. For example, a reform implemented in 1992 neglected and reduced refining capacity. The enduring challenge of the state's fiscal dependency on PEMEX was never remedied. This took a toll on the company's capacity to invest in projects that could ensure its long-term viability. As systemic corruption continued unchecked, the situation would start becoming critical by the mid-2000s.

Mexico's oil production has been steadily decreasing since 2005 because of the natural decline of its main oilfields. In a constitutional reform approved in December 2013 the Mexican state finalized the 75-year monopoly of PEMEX, ending the control the company had over every segment of the energy sector (exploration, production, transportation, refining and commercialization). Despite this liberalization, PEMEX remained the Mexican state's stakeholder in the sector.

Regardless of the 2013 reform effort, structural shortcomings leading to declining capacity were far-reaching and persisted four years after the new legislation's approval. In 2017, Mexican oil production fell 42% compared to the production peak reached in 2004 of 3.3 million barrels per day (bpd) (Deloitte 2014). In 2003, PEMEX was the third oil producer in the world, but by 2017 it had fallen to the twelfth spot (Oxford Business Group 2018). This situation can be explained partly because, regardless of the liberalization of reserves to exploration and exploitation by private investors, these projects are not yet at the production stage, leaving most production under PEMEX's control.

These circumstances signalled a wider crisis within PEMEX involving rampant corruption and long-existing inefficiencies. For example, in 2016 PEMEX reported losses equivalent to $17.7 billion USD, a 74% increase compared to 2015 (*El País* 2018). This situation was exacerbated by the fall of global oil prices in mid-2014. In 2015 the price of the Mexican barrel decreased from $86 to $43.3 USD. In January 2016 it reached $18.90 USD (Obregón 2016). As Lajous (2018) summarized, in 2018 PEMEX was facing a contradiction posed by the reform discourse of ambitious promises and the reality of a company operating on the verge of collapse. Within this dichotomy, aspirations for private investment and modernization collided with the realities of falling energy reserves, decreasing refining capacity, inefficiency and pervasive corruption.

Losses coupled with collapsing oil prices became unsustainable by 2016, when a cut of a $100 billion MXN ($5.3 billion USD) to PEMEX's budget was announced, an annual reduction of 21%. This translated into a fall in production of 2.52 million bpd in 2013 to 1.85 million bpd in 2017 (-36%) (Siglar 2018). 11,615 PEMEX workers were laid off as a part of the 2016 cut and approximately 28% of PEMEX's workforce lost their jobs during the EPN administration (Miranda 2018). By January 2017 shortages started occurring, as declining capacity combined with rising demand (Fitch Solutions 2019, 45). Meanwhile, this energy crisis was unfolding within Mexico's deteriorating security situation.

The Security Crisis

Mexico's hegemonic party the National Revolutionary Party (PRI) had historically established corrupt agreements with few predominant criminal groups, mostly drug trafficking organizations. This brought stability, as centralization in the political and the criminal realms were key for the prevailing *Pax Narca*. Lack of competition in both instances allowed for leadership continuity, permitting agreements to be sustained. The presence of the PRI at all levels of government established command chains that allowed it to diffuse local conflicts: if criminal disputes emerged, state governors would coordinate with central authorities to neutralize them. These mechanisms regulated the relationship between criminals and the state keeping violence levels low (Morris 2013, 207). At the centre of this arrangement, security functions were concentrated by the Attorney General (PGR). These elements allowed the state to maintain its monopoly over power despite having inefficient and underdeveloped police forces (Durán-Martínez 2018, 94). The key was in the state security apparatus' ability to pose a threat of coordinated response, presenting a significant risk to criminals.

These arrangements would start to erode by the 1980s–1990s as Mexico's democratization process advanced inconspicuously. Through the 1980s, the opposition party *Partido Acción Nacional* (PAN) began to claim victories in municipal elections, particularly in Baja California, Chihuahua and Sonora. The municipal election victories soon gave way for changes at the state level: the PAN won the governorships of Baja California in 1989, Chihuahua and Guanajuato in 1991 and Nuevo León in 1997. In 1997 for the first time in the country's history, the PRI lost its majority in the legislative lower chamber. This process would culminate in the ousting of the PRI from the presidency in 2000. Electoral liberalization occurred in parallel to institutional changes. The once all-powerful PGR underwent various reforms between the 1980s and 2000s fragmenting the security apparatus. The increased political pluralism battered the presence of a dominant political entity that could sustain stabilizing agreements. As proof of this, by 2007–2012 only 10% of municipalities had a mayor with the same political affiliation of the Governor and the President (Trejo and Ley 2016, 31). The centralization behind the law of rules was vanishing.

The pluralization of Mexico's political system and its tendency towards increased competition also occurred in the criminal underworld. The collapse of the Guadalajara Cartel in the 1980s led the formation of the Sinaloa, Tijuana and Chihuahua organizations. These reconfigurations increased the criminal organizations involved, enhanced competition and generated confrontations. Seizures, arrests, and armed clashes grew, further increasing fragmentation since the mid-1990s. Yet, violence remained largely hidden because the arrangements of the *Pax Narca* still lingeried. It is important to clarify that not all changes leading to Mexico's security deterioration were driven by internal forces, as the US also played a major role. The American closure of the route of the Caribbean made Mexican drug cartels the new key players of regional cocaine smuggling, which led to a considerable increase in their profits and capacities. Increased criminal power combined with growing enforcement, fragmentation and increasing competition.

An event occurred during the Zedillo presidency (1994–2000) that would reshape Mexico's criminal underworld. The power of the Gulf Cartel (GC) grew dramatically during that administration. This, combined with increasing international pressure, led to the capture of the GC leader Juan García Ábrego in 1996. Before this, García Ábrego had been protected by high-ranking officials. His arrest signalled that protection for criminal leaderships had become unstable and less dependable. The increased uncertainty led Ábrego's successor, Osiel Cárdenas, to form in 1997 a paramilitary arm of the GC: the Zetas. The group was formed by former elite soldiers. Yet this group was not solely formed as a response to the erosion of state agreements with criminal groups, criminal dynamics played a role. The murder of Salvador Gómez, another leader of the GC who shared the control of the organization with Cárdenas, also influenced the foundation of the Zetas as a GC armed wing; strengthening security and cohesion to an organization destabilized by in-fighting (Atuesta and Pérez-Dávila 2018, 2).

The Zetas subsequently proved formidable in stabilizing the GC and expanding its violence capacity. Their growing power, pushed by extreme violence, aggressive expansion and military professionalism, heightened criminal competition during the early 2000s. Consequently, new criminal groups without the institutional and financial collusion of older criminal organizations started to appear. These groups had a greater tendency towards violence than their more consolidated counterparts who maintained institutional agreements (Morris 2013, 209). The Zetas were rewriting the rules of the game in Mexico's criminal underworld and their centrality to the deterioration of public security cannot be stressed enough. They influenced the change in the nation's criminal violence towards more visibility, brutality and increased diversification towards new black markets (Durán-Martínez 2018, 105).

The Zetas' irruption in an already unstable context, made criminal groups more territorial and violent, all while the state lost its regulation capacity over the criminal realm. By June 2005, violence in the northern states was increasing. In response, the federal government deployed a military intervention in Baja California, Sinaloa, Tamaulipas, Coahuila, Sonora, Chihuahua and Veracruz. These tendencies reached a crescendo in 2007, when newly elected President Felipe Calderón launched a "war against drug trafficking", targeting all sort of criminal leaderships and increasing fragmentation. Regional intervention of the military and federal police increased the loss of state cohesion by displacing local and municipal enforcers, eroding protection agreements and communication channels between local authorities and criminals. This combination turned Mexico into a powder keg: unconditional crackdowns by the authorities, coupled with strong criminal organizations and turf wars, led criminals to use violence increasingly as a deterrence tool to coerce opponents and make enforcement agents change their behaviour towards them (Lessing 2015).

Nevertheless, it is important to reiterate that the growth in homicides was not abrupt. The state security apparatus' erosion was gradual, driven by changes in the political system and security policy fluctuations. From the 1980s to the mid-2000s Mexico remained a peaceful country. In 1980 and 1990 the national homicide rate was 18 for every 100,000 inhabitants. By 2000 it had decreased to 11 and in 2005 it reached 9.4 (Menéndez 2012, 179). This could be explained because mechanisms for managing security were deep-rooted and their disappearance was a consequence of long-running processes.

Indiscriminate state enforcement against drug trafficking eliminated any incentive for criminals to keep their operations hidden or to abstain from violence. The calderonista crackdown also made evident the Mexican state's coercive capacity, showing considerable shortcomings which encouraged criminals to challenge security institutions. The limitations of militarized repression reinforced some criminal groups while weakening others, spawning the aforementioned smaller and more violent organizations.

Changes affected corruption. Democratization made the political system more diverse. In modern-day Mexico higher-level officials have lost most of their capacity to control their lower-level counterparts. Mexican criminal groups are now facing a more pluralistic and decentralized state in which the variety of corruptible agents multiplied. Corruption became more contested amongst criminals and directing violence at state representatives colluded with rival organizations increased attacks against authorities (Morris 2013, 209). In this context of increased criminal competition, political decentralization and all-encompassing state repression, criminal groups adopted violence as a dominant strategy for survival (against other criminals and authorities). As a result, Calderón left office with a legacy of over 102,000 people killed in 2012.

The Peña Nieto Years

In 2012 EPN's presidency shifted attention from security to reforming Mexico's economy. Yet a militarized security strategy remained, now under a reinforced Interior Ministry relying on "enhanced institutional coordination" to improve the dismal security situation. This diagnosis was influenced by a nostalgia of the hegemonic era and the conviction that the state had the tools to improve security. Under this logic, fragmentation was a matter of political skill, not of deep-rooted structural trends.

This analysis was backed by a homicide decrease. Yet local cases point to more complex dynamics than a uniform violence reduction. In Ciudad Juárez and other northern regions, violence decreased (Durán-Martínez 2018, 106). In 2010 nearly 25% of all homicides in Mexico (25,757) took place in Chihuahua (6407, of which 3766 occurred in Ciudad Juárez). The municipality of Ciudad Juárez represented 15% of homicides nationwide. Between 2010 and 2014 homicides in Ciudad Juárez declined 84%. Decreases also occurred in Monterrey (77%) and Torreón (75%). If the decreases in Chihuahua, Coahuila and Durango are put together they justify the bulk of this reduction (Fig. 3.1).

If these states are deducted from national homicides, violence was prevalent elsewhere. In Guerrero and Michoacán violence was escalating.

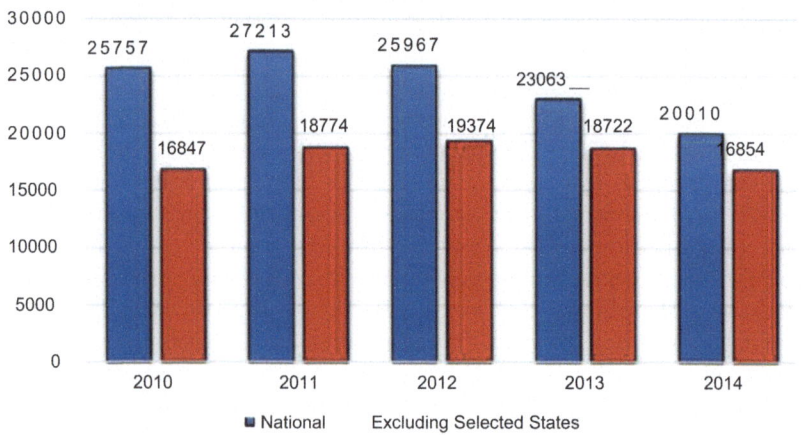

Fig. 3.1 National homicides vs totals excluding Chihuahua, Coahuila and Durango 2010–2014 (Hope 2017)

Consequently, in both states paramilitary groups were formed against criminal organizations (Durán-Martínez 2018, 106). In 2015 homicides started rising again. By 2016 it was clear that violence was a growing nationwide trend. By October 2016, intentional homicide rates tripled in Colima, increased 161% in Veracruz, 93% in Zacatecas and 62% in Michoacán. During that period, homicides grew in 23 of Mexico's 32 states (Hope 2017).

To what can the early decrease in homicidal criminal violence be attributed to? According to the Brookings Institution, the main factor influencing these reductions (without neglecting police surveillance coordinated with socio-economic programmes) is a modern version of *Pax Narca* in which the violence decreases are achieved through the control that a criminal group manages to establish *"over a particular territory and its corruption networks"* (Felbab-Brown 2019, 4). It is through this control that a dominant group can deter potential competitors *"and prevent violent confrontation with rivals"* (ibid.). Other researchers also recognize this dynamic (Fuerte Celis et al. 2018, 16). An example of this scenario is the decrease in violence experienced in Ciudad Juárez once the Sinaloa Cartel became the predominant criminal network in that city. This argument warns us about the state's limitations in its influence over criminal violence.

The worsening security situation during EPN's term exhibited that local dynamics mattered in determining interactions between the state and criminals, while the failure of improving institutional coordination by the presidency showed the days of hegemonic stability were gone and reconfirmed the state's security limitations. Aspirations to return to the days of few dominant organizations and the hegemonic peace proved delusional, there *"were too many interests and power centres for that to be possible"* (Durán-Martínez 2018, 107). The growth of violence reached a dismal pinnacle in 2017, when 31,174 people in Mexico were murdered, increasing the national murder rate from 20 to 25 people killed per 100,000 inhabitants between 2016 and 2017 (INEGI 2018). The growth of violence and its territorial dispersion showed a diversification trend of criminal networks towards exploiting local black markets, fuel trafficking amongst them. Diversification to this illicit market expanded violence to places with no prior criminal history.

EPN's presidency ended with over 120,000 homicides (Aguilar 2018). In this context of sustained deterioration of the energy sector, rising violence, fragmentation, criminal diversification and unruly corruption, the MFBM would find ground to consolidate. As it was explained, corruption has been a part of PEMEX since its inception as a SOE. But the levels reached during the EPN administration were shocking. Besides the multilevel co-option that enabled the expansion of fuel trafficking, there is evidence that the Brazilian company Odebrecht secured contracts through bribes directly implicating former PEMEX Director Emilio Lozoya (Morales 2020, 17). To understand how the consolidation process of the MFBM began it is important to explore the origins of fuel trafficking and explain how it became a priority for criminal networks. Here, the case of the Burgos Basin comes into play as the genesis of fuel trafficking as a relevant illicit enterprise.

The Origins of Mexican Fuel Trafficking

Different sources mention that fuel trafficking in Mexico was first noticed in the late 1990s. Maritime fuel trafficking was first detected by auditors during the presidency of Carlos Salinas (1988–1994). Back then, a warning was made about the lack of monitoring in PEMEX's Marine Terminals, where large volumes of hydrocarbons were transported (Pérez 2018c). The first official sources mentioning fuel-theft in Mexico as a concern date back to 1997, until then small-scale fuel-theft was the defining characteristic (Pérez 2018a).

Between 1996 and 2001 the amount of illegal extraction points found in pipelines by PEMEX pointed to a downward trend (Fig. 3.2).

Still, PEMEX conveyed that it prioritized tackling fuel-theft, formula adulteration and unregulated imports (Córdoba 2003). In 2004 PEMEX made public that between 1998 and 2002 it suffered losses of $16 billion MXN ($1.6 billion USD) related to theft, adulteration and fuel smuggling (Vicenteño 2004a). Back then it was pointed out that in Hidalgo's Tula refinery, seven investigations were opened because of inconsistencies between tanker trucks leaving that installation and their reported cargo. The same fuel-theft method was detected in the Storage and Distribution Terminal of Zapopan, Jalisco. Tanker truck drivers also declared that colleagues were being targeted by criminals in Mexico's highways, having their vehicles stolen from them to rob jet-fuel, gasoline and diesel (ibid.).

In 2004 the federal government launched the programme "*Combating the Illegal Fuel Market*" (CMIC). Memorandums of a 2007 CMIC meeting mentioned that, historically, national fuel consumption had a behaviour corresponding to GDP growth. This situation changed: by

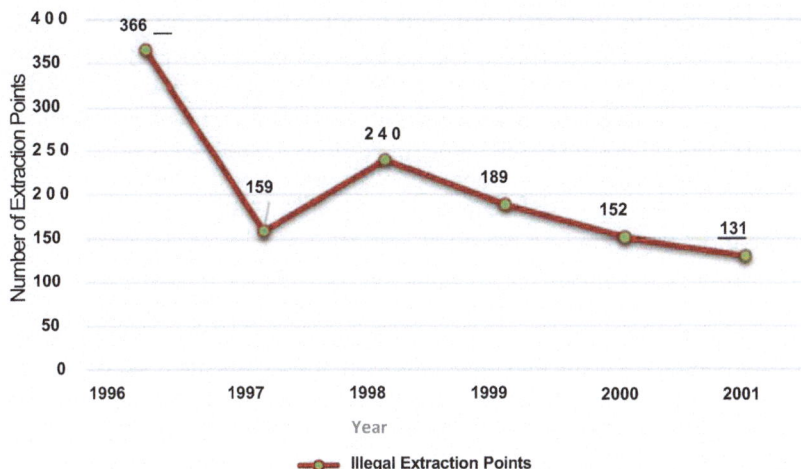

Fig. 3.2 Extraction points PEMEX pipelines 1996–2001 (Córdoba 2003; Cervantes 2012)

1997 national fuel sales decreased compared to GDP growth. This reduction was attributed to a "*growing illicit fuel market*" that represented losses between $4.5 and $6 billion USDbetween 1998 and 2002 (a daily loss of 52,000 barrels) (Cervantes 2012; "Memoria de Labores 2008", 113).

By the mid-2000s reports pointed to the existence of sophisticated fuel-theft and distribution operations in Mexico that included the processing of large amounts of stolen fuels, professional installations and equipment protected by illegality camouflaging. In 2004 the Federal Police and PEMEX Refining (a subsidiary) detected fuel-theft groups in Tlaxcala, Estado de México, Oaxaca, Tamaulipas, Querétaro and Sinaloa. During a police raid in Estado de México, a fuel trafficking operation was exposed in which four containers, holding 500,000 litres of stolen derivatives each, were seized. The same sight of this discovery had the equipment required for pumping and loading fuel to tank trucks operating under a façade company known as "*Transportes de Petroquímicos SA de CV*" (Vicenteño 2004b).

In 2005, journalistic sources cited security reports warning about eight fuel trafficking networks that operated in the west, north and centre regions of Mexico. These groups relied on specialized structures with cells dedicated to fuel extraction and façade companies for distribution. Two individuals belonging to one of these networks had 17 gas stations with operating permits granted by PEMEX. This same network owned a cobblestone factory, a manufacturing enterprise and two transport companies to conceal its fuel trafficking operations taking place in multiple states. Adding to this operational sophistication was the fact that one of its members was a PEMEX employee (Medellín 2005). It is important to mention that none of the cited sources connects these groups with large criminal organizations. Despite the complexity of their operations and structures, it seems these networks operated with a certain degree of autonomy. During these years, the MFBM went through a growth period: in 2006 PEMEX estimated that fuel-theft losses equalled $4.39 billion MXN ($405.8 million USD); by 2007 it increased to $4.75 billion MXN ($435 million USD) and in 2008 it reached $5.8 billion MXN ($422 million USD) (Hernández and Córdoba 2009).

Pipeline-theft remained constant during Vicente Fox's (2000–2006) administration: in 1999 189 extraction points were detected, 152 in 2000, 131 in 2001, 155 in 2002, 152 in 2003, 110 in 2004 and 136 in 2005. Felipe Calderón's administration marks the beginning of pipeline fuel-theft growth: in 2006, 220 shots were discovered; in 2007, the launch of the "war on drug trafficking", 323. By 2008, 396 were found, in 2009, 439 and in 2010, 631 ("Memoria de Labores 2009", 137; "Memoria de Labores 2011", 131; Cervantes 2012). Between 2007 and 2008 the largest increase in the number of pipeline illegal extraction points detected in the history of PEMEX took place, with a growth of 22.6%. By 2011, discovered extraction points more than doubled to 1324 (EnergeA and Grupo Atalaya 2017). These numbers indicate that, under Calderón, criminal networks diversified towards a growing MFBM.

According to former PEMEX Director Jesús Reyes Heroles, at one point the most stolen fuels in pipelines were diesel and gasolines, but this changed. Since 2008 the SOE detected an increase in the theft of gas condensate. Between July 2008 and July 2009, it was estimated that $5.8 billion MXN ($442 million USD) worth of gasolines, diesel and other derivatives were lost to the MFBM. During that period, $3.5 billion MXN ($265 million USD) worth of condensate gas was lost (Hernández and Córdoba 2009). And this takes us to the Burgos Basin, the place where the first case of large-scale criminal involvement in fuel trafficking was exposed.

The Río Grande Valley Revelation

Río Grande Valley is a lush floodplain in southern Texas, a picturesque area that attracts tourists from the northern US, Canada and Mexico. In the spring of 2007, an unexpectedly significant event would occur in that place. The Texas Department of Public Safety detained a truck with $1,149,069 USD driving to the Río Grande Valley (Garay-Salamanca and Salcedo-Albarán 2016, 100). The authorities seized the money. Later, a petition was made by the owner to have the money returned, a man named Luis Rivera Rodríguez. The claim was backed by 1390 documents demonstrating the lawfulness of those resources. These papers included invoices and permits supporting Mr. Rivera's entitlement to commercialize Mexican oil derivatives, the source activity for the seized cash.

Everything about the documentation seemed legal. But two other institutions got a hold of them: The Drug Enforcement Agency (DEA) and the U.S. Immigration and Customs Enforcement (ICE). These institutions began exchanging information and noticed a detail: the only agent lawfully permitted to sell oil derivatives in Mexico was PEMEX. What would follow would be the uncovering of the first large-scale fuel trafficking operation executed by criminal networks between Mexico and the US. With this discovery the existence of a recent and relevant criminal black market would be exposed.

The Burgos Basin Operation

The impossibility of Rivera trading fuels pointed to an illegal operation. American authorities proceeded to investigate a group of energy companies that were reporting extraordinarily high revenues importing Mexican fuels. The Río Grande revelation would have extraordinary implications. By July 2008, ICE and the DEA discovered that Sun Petroco LLC (Rivera Rodríguez's company) was selling diesel and gas condensate in Texas at ¢35 a gallon (Garay-Salamanca and Salcedo-Albarán 2016, 101), well below the market value of $3.8 USD (EIA). One Brownsville company also stood out in its volumes of acquired hydrocarbons: Continental Fuels.

Another detail authorities uncovered was that custom agents were involved, certifying hydrocarbons in Mexico as organic waste for export. Once on US soil, tanker trucks rebranded their cargo as fuel. In

August 2008, a suspicious condensate operation involving the companies Continental Fuels, Petro Source, Transmontaigne and a transport company linked to Rivera Rodríguez caught the attention of American law enforcers. By September 2008 Rivera Rodríguez was detained on charges of fuel-theft for exporting into the US and money-laundering of drug trafficking resources (Pérez-Treviño 2008).

Once in custody, Rivera Rodríguez confessed participating in a large fuel trafficking operation taking place in Mexico's Burgos Basin since 1998 (corroborating the information of 2007s CMIC programme). He revealed that entrepreneurs, financial operatives (Pérez 2011, 53–61) and politicians were involved in a criminal network with the Gulf Cartel (GC) and the Zetas specialized in fuel smuggling. All the Zetas and GC leaders received revenues from fuel trafficking into the US. Rivera Rodríguez would pay $80,000 USD for crossing four tanker trucks daily and each truck generated profits of $15,000 USD ($1.8 million USD monthly). Besides condensate trafficking operations included naphtha, diesel and other derivatives.

The Burgos Basin is a shale reserve in north-eastern Mexico. It covers 70,000 km^2 and in 2011 had 2187 hydrocarbon-producing wells spread across 98 municipalities in Tamaulipas, Nuevo León and Coahuila. It is the most important non-petroleum gas reserve in Mexico, and it is estimated to concentrate two-thirds of the country's recoverable shale gas resources. At the time, the hydrocarbons extracted were transported to 150 collection stations and 52 transfer centres. The condensate was transported to collection centres by tanker truck. Once there, they were taken to the processing complex of PEMEX in Reynosa (yellow dots on map). Another part of the production was sent to the Cadereyta Refinery in Nuevo León (red dot on map) (Map 3.1).

According to Rivera's testimony fuel trafficking operated as follows (Illustration 3.1).

As investigations continued, the implications would reach the commanding heights of American politics. The connection was made through the Vice President of Continental Fuels, Joshua Crescenzi. US authorities took notice of the unusually high profit margins this company had shown since 2006. This company, with Crescenzi as an alleged relevant operator,[1] received in its heyday 6000 barrels of stolen Mexican

---

[1] Crescenzi's participation in this case was mentioned by multiple involved parties. In a Dispositive Motion, *Pemex Exploración y Producción* (PEP) accused Continental Fuels and

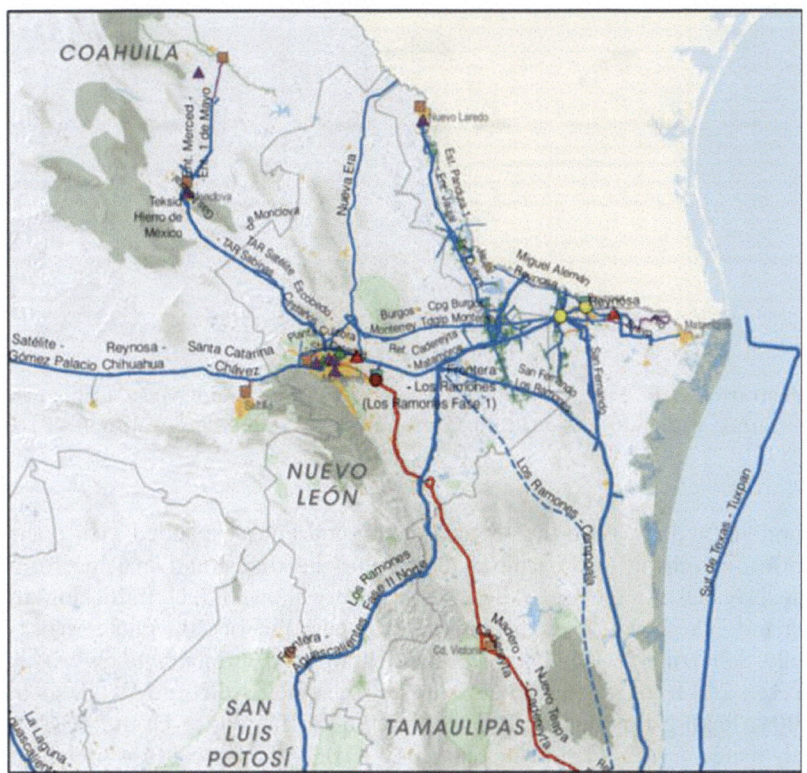

Map 3.1 Energy infrastructure in the Burgos Basin (Llano and Flores 2017)

Joshua Crescenzi of selling stolen Mexican condensate citing the deposition of Timothy Brink, president of Continental Fuels. Brink testified that in July of 2008, Josh Crescenzi (his employee at the time, acting as vice president of operations for Continental Fuels), told him that the product he was purchasing from Mexico was stolen. Brink also testified that, once he analyzed paperwork documenting shipments of product that he bought from Mexico, he detected discrepancies that revealed its illegal origins. PEP also cited the guilty plea of an individual named Arnoldo Maldonado, owner of the company Y Gas & Oil. According to this source, Y Gas & Oil supplied condensate stolen in Mexico to Continental Fuels. Another source cited by PEP was the guilty plea and the deposition of Trammo's president, Donald Schroeder. In these documents Schroeder declared that Continental Fuels was dealing condensate stolen in Mexico and that his company, Trammo

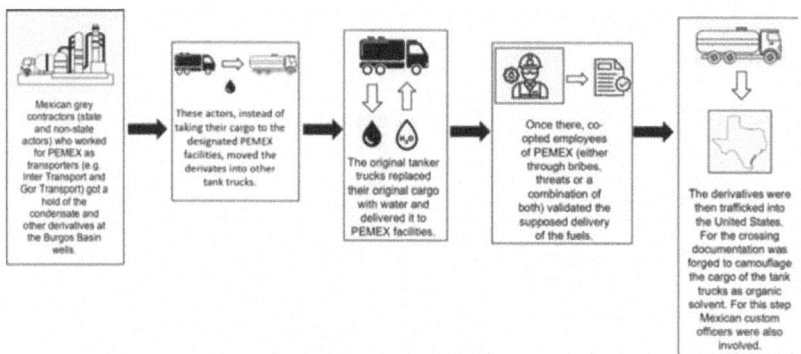

Illustration 3.1 Fuel trafficking scheme in the Burgos Basin. *Source* Elaborated by author with information from Garay-Salamanca and Salcedo-Albarán 2016, 96–110

condensate daily. With the profits Continental Fuels expanded their operations, acquiring port facilities in Brownsville that could manage from small vessels to tankers. These acquisitions allowed fuel trafficking to expand. Crescenzi allegedly provided an average of 200 tanker trucks daily, delivering 20,000 barrels of condensate to Continental Fuels weekly (Pérez 2011, 64). Crescenzi was press liaison for George W. Bush in 2004, linking the Burgos Basin operation to the White House (Garay-Salamanca and Salcedo-Albarán 2016, 105). Crescenzi was detained and became a protected witness. With his arrest the far-reaching implications of this criminal enterprise became clearer. Big companies like German-based BASF were involved, as well as Royal Dutch Shell.

Petroleum, sold this condensate to larger companies within the United States including BASF.
Sources:
PEP's Dispositive Motion, Docket Entry No. 492, p. 4 (citing Deposition of Timothy Brink, Exhibit 10, Docket Entry No. 492-12, pp. 13–14).
Plaintiff PEP's Reply in Support of Motion [Dkt. 492] ("PEP's Reply in Support of Motion"), Docket Entry No. 577, pp. 15–16.
PEP's Dispositive Motion, Docket Entry No. 492 (citing Transcript of Rearraignment, Exhibit 12, Docket Entry No. 492-14, p. 20).
PEP's Dispositive Motion, Docket Entry No. 492, p. 5 (citing guilty plea of Donald Schroeder, Exhibit 14-72 in Docket Entry No. 492, pp. 19–21).

Regarding BASF FINA, it was revealed during court proceedings that it acquired stolen condensate from the distributor Trammo Petroleum. Trammo admitted buying fuels while aware of their illicit origin. BASF FINA processed the stolen condensate in Port Arthur, Texas (Reinhart 2014, 770). Donald Schroeder, former president of Trammo Petroleum, gave further details about the trafficking operation coming from the Burgos Basin, corroborating the scheme exposed by Rivera Rodríguez: American import companies acquired condensate stolen from PEMEX to sell it to larger corporations like BASF and Continental Fuels. Companies even transported stolen fuels in their tanker trucks across the border (ibid., 768–769).

A truck driving to the Río Grande Valley uncovered a conspiracy that reached the highest levels of political and business power. It exposed a transnational criminal network, revealing that Mexican fuel trafficking had become a relevant black market for large criminal groups. But the repercussions of this case in Mexico are still to be explored.

Burgos: The Mexican Side

The GC/Zetas criminal network co-opted PEMEX since the mid-2000s. Clashes between PEMEX's Physical Security Services unit (GSSF) and the GC were reported since 2006. The GSSF documented how fifteen GC *sicarios* threatened their personnel in Reynosa, Tamaulipas. During the incident, the weaponry of the GSSF was taken by the criminals who offered them $1000 USD in bribes. Companies operating in the Burgos Basin blamed all theft incidents on their drivers. When caught, they received legal assistance from lawyers who defended PEMEX workers involved in fuel trafficking. With grey actors linking criminality to PEMEX, fuel trafficking expanded. Extraction in wells started being complemented with increased tanker truck theft targeting PEMEX and its contractors.

Armed commandos operated all across the Burgos Basin. Their lookout networks (*halcones*) would notify them whenever a tanker truck was nearby to steal it. These groups had access to PEMEX's internal communication system, which required specialized equipment and access codes. Some examples show the ferocity these groups displayed when it came to their fuel trafficking operations. On the night of July 23rd, 2007, in the Reynosa-San Fernando Highway a tanker truck was detained with 30,000 litres of stolen condensate. After being stopped eight pick-up trucks carrying armed commandos surrounded the agents, identifying themselves as members of the GC. The commando leader offered $50,000 MXN to the agents. They did not accept the money and left being outgunned by the criminals. The GC and the Zetas responded aggressively to the decommission of tanker trucks, kidnapping PEMEX workers in response and sending heavily armed groups to retrieve vehicles from storage facilities (Pérez 2011, 58).

As American demand for stolen fuels grew, criminal groups started taking over wells. In 2010, Zetas armed cells took over the compression station *Gigante-1* in Nueva Ciudad Guerrero, a municipality close to Reynosa (Correa-Cabrera 2017, 198). Forty armed men riding four pick-up trucks restrained the workers at that station and kidnapped six of them. An insider source explained the dreadful situation of the Burgos Basin: "*'What they want is total control of the area. They have launched threats that they will kidnap anyone who shows up to work in that plant',* said the source, who acknowledged that for years these groups have stolen fuel from more than a thousand wells in the Burgos Basin*" (*Proceso* 2010). Targeting PEMEX workers and its contractors increased when the GC and the Zetas fragmented in 2010. That year, sixteen PEMEX employees were kidnapped across the Burgos Basin (Correa-Cabrera 2017, 197).

After the Zetas separated from the GC fuel trafficking operations continued, with trucks being marked with each organization's brand. Trafficking into the US took place through legal ports of entry (Reynosa, Nuevo Laredo and Tampico). Other two entries were the private terminals of Port Arthur and Port Isabel in Texas. Companies like Transmontaigne contributed with storage to triangle stolen condensate. As aforementioned, private trucks were used for trafficking: in 2007 the Mexican Army found fourteen trucks filled with stolen condensate belonging to private businesses.

An articulating figure behind this operation was Miguel Ángel Almaraz Maldonado, a politician and businessman from Tamaulipas. He owned companies specialized in distribution and commercialization of fuel in the US and had a career in the Democratic Revolution Party (PRD). Maldonado consolidated his power when he financially supported the presidential campaign of AMLO in 2006 (Garay-Salamanca and Salcedo-Albarán 2016, 111–112). He occupied the PRD leadership in Tamaulipas from 2005 to 2007 and, according to Rivera Rodríguez and other witnesses, was the main trafficking operator on the Mexican side of Burgos. He reportedly paid $600,000 USD a month to Jaime González

Durán *El Hummer*, a Zetas leader, to carry out his fuel trafficking activities and had a close relationship with Alfonso Lam *El Gordo Lam*, a GC leader.

As previously mentioned, the Gulf-Zetas' leaders profited from fuel trafficking. The dynamics behind these payments and the role of leaderships within this criminal network are noteworthy: according to Rivera Rodríguez, leaders of both groups acted swiftly whenever they didn't receive their fuel trafficking payments. Rivera was once kidnapped by Miguel Ángel Treviño *El Zeta 40* for owing $2.7 million USD. In another incident, when he went to pay his monthly share, Rivera saw a friend of his tied down to a chair for having taken stolen fuel belonging to another criminal leader. This modus operandi illustrates Kenney's (2007) argument that one of the main functions of leaders within criminal networks is solving disputes amongst participants (243).

Almaraz Maldonado is an example of a grey actor who, using his presence and resources in both politics and business, provided key resources for the network to expand their fuel trafficking operations. He bribed public officials, provided tanker trucks and also laundered money by coordinating 150 bank accounts for this criminal operation. This actor thus provided the Gulf-Zetas network with important legal and illegal resources allowing the development and expansion of their criminal enterprise. US authorities informed their Mexican counterparts of this operation in 2007. Almaraz Maldonado would not be detained until 2009.

These operations turned out to be a success. By 2007, according to PEMEX, 40% of the hydrocarbons produced in Burgos were controlled by criminals (Garay-Salamanca and Salcedo-Albarán 2016, 138). Astonishingly, PEMEX had forty security employees in the space covered by the Burgos Basin between 2007 and 2009, an area the size of Ireland (Pawley 2014, 144). This lack of personnel rendered long-distance monitoring systems useless, as there was no capacity of sending people to remote areas where subtraction incidents were detected (Garay-Salamanca and Salcedo-Albarán 2016, 110). During 2007 over 51% of all the stolen hydrocarbons seized by PEMEX was condensate followed by crude oil with 18.6% ("Memoria de Labores 2007", 96). This shows the importance condensate (a high-quality light oil) had for the black market that supplied American refining companies.

From 2007 to March 2009, the Zetas-Gulf network trafficked 175,855,251 litres of hydrocarbons and more than 3519 tonnes of oil derivatives across the border. PEMEX's inquiries estimated the losses in $508,548,320 MXN, which adjusted for inflation represent a loss of $1.5 billion MXN ($79.4 million USD) (Garay-Salamanca and Salcedo-Albarán 2016, 110). Beyond these costs, the significance of the Burgos case was greater. Its discovery marks the beginning of large criminal networks diversifying into fuel trafficking; a costly price to pay because of ineffective enforcement and another threat to an embattled PEMEX.

It marked a pivotal moment for the Zetas becoming a more sophisticated criminal network, implementing a diversification model that gave them a transnational scope. This reach allowed them to engage in large-scale fuel trafficking and launder revenues establishing companies that became PEMEX contactors. The Burgos Basin represented the genesis for Mexican large-scale fuel trafficking and would set the stage for it to become one of the main national black markets as criminal diversification grew, the co-option of PEMEX and other grey actors progressed unchecked, and hydrocarbons became a more profitable illicit enterprise. Eventually the Zetas expanded their fuel trafficking operations further within Mexico.

The geography of pipeline extraction points can give us a glimpse as to how this modality of theft drifted during the 2000s. This in turn can give some clarity as to how the MFBM behaved geographically during that period. The School of Government of *TEC de Monterrey* compiled 44,237 illegal extraction points across Mexico between January 2003 and August 2018 (Carranza 2018). With this information they put together maps that show us the following phenomena: between 2003 and 2006 the states of Coahuila and Veracruz were pipeline fuel-theft hotspots. By 2007–2010 pipeline theft consolidated in the northern states of Coahuila, Nuevo León and Tamaulipas, while Veracruz remained a hotspot. By 2009–2010 pipeline theft operations start showing a stronger presence in the centre areas of Mexico. Between 2011 and 2014, Nuevo León, Tamaulipas, Sinaloa, Guanajuato and Estado de México appear as pipeline theft hotspots at a time when the MFBM becomes a criminal enterprise present across Mexico's territory, particularly in the Gulf Coast. Finally, in 2015–2018 pipeline theft integrated the centre-Bajío regions as key areas for this criminal market.

Having explained the energy and security crises, the beginning of fuel trafficking as a relevant criminal enterprise and its dispersion, it is time to analyse the astonishing growth this black market achieved, and the criminal actors involved.

Explaining the MFBM

The MFBM is a profitable enterprise that has become a priority for criminal networks of all capacities. Using the typologies of Beckert and Dewey (2017), we can classify the MFBM as a Stolen Products type 2 black market that uses invisible supply chains for commercialization in legal and illegal retail points. The MFBM also has offshoots that make it a hybrid illicit market in which hydrocarbon theft is combined with fuel adulteration, making it fit under the Counterfeits type 3 category (Pérez 2011, 39–40; Notimex 2015). Finally, this market also falls under the classification of Violations of Regulations type 5 market.

Another characteristic of the MFBM, based on my findings, is its reliance mostly on demand that has grown and consolidated within Mexico. Supply for the analysed period mostly exploits internal energy installations but cases of transnational smuggling coming into and from Mexico have also been found. Counterfeit formulas do exist, but they do not represent a substantial portion of the market which is dominated by original formulas sold through lawful and unlawful points of retail. One example of how the counterfeits side of the MFBM combined with transnational smuggling appeared in 2000. At that time, Ana Lilia Pérez (2011) revealed that transnational fuel smuggling of different fuels and solvents were entering Mexico from its northern border. These derivatives were used to make counterfeit fuels that were commercialized through PEMEX service stations (39–40).

The Relevance of the MFBM

Mexican fuel trafficking showed an extraordinary growth between 2011 and 2018. This expansion manifested itself in the increase in illegal pipeline extraction points (Fig. 3.3).

Illegal extraction points estimate the growth of the MFBM indirectly, whereas volumetric losses offer more accurate data. According to internal data of Mexico's energy regulator, between 2011 and 2015, roughly 50,245,509 barrels of hydrocarbons were stolen in Mexico (7,989,035,931 litres) through 15,847 illegal extraction points installed across different pipelines in the country (EnergeA and Grupo Atalaya 2017, 30). The problem between 2016 and 2018 grew alarmingly. The number of illegal extraction points reported during 2018, 2017 and 2016 (14,894, 10,363 and 6873 respectively) add up to a total of 32,130 illicit

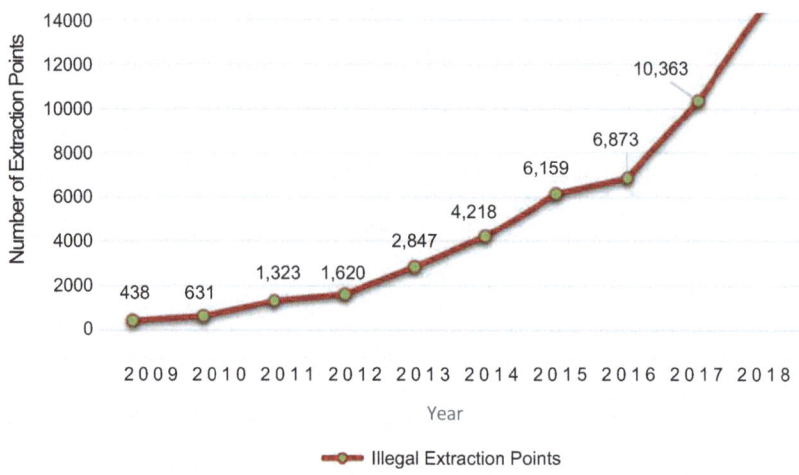

Fig. 3.3 Extraction points PEMEX pipeline network (PEMEX: "Reporte de tomas clandestinas"; EnergeA and Grupo Atalaya 2017, 344; Transparency Petition Response 1857000059819)

tappings. This represents a 90% increase of extraction points in 3 years, when compared to the 4-year period between 2011 and 2015.

An estimate assessed fuel-theft losses in PEMEX's pipeline network at $160 billion MXN ($7.97 billion USD) between 2009 and 2016 (Montalvo 2017). This figure is double the annual budget for *Prospera* (ibid.), the main poverty alleviation programme in Mexico, and is equivalent to the original investment required for Mexico City's cancelled international airport (Villamil 2018). This estimate was conservative, since it was elaborated assuming that the fuel stolen from pipelines (polyducts) was only the cheapest gasoline in Mexico. Since polyducts transport other higher-priced derivatives (diesel, *Premium* gasoline) this figure was certainly higher. Pipeline extraction is complemented with robbery of tank trucks, storage facilities, refineries and marine terminals. In its 2017 Report, PEMEX identified the "*Illicit Fuels Market*" as a meaningful threat to the company's objectives ("*Informe Anual 2017*", 2018b, 74). Up to this point this was the information available to assess the MFBM. This was about to change with the beginning of Obrador's presidency.

The newly installed President declared in December 2018 that fuel trafficking represented losses of $50–$70 billion MXN a year ($2.5–$3.5

billion USD) (Raziel 2018). Subsequently, more detailed data was made public confirming the size of the MFBM: in 2016 9,490,000 fuel barrels were stolen, representing a loss of $30 billion MXN ($1.5 billion USD) for PEMEX. In 2017 15,695,000 barrels were stolen worth over $51 billion MXN ($2.6 billion USD). During 2018, 21,243,000 barrels of fuels were stolen, a loss of $66 billion MXN ($3.4 billion dollars) (López Obrador 2018). Approximately 9 million litres of fuel were stolen daily, the equivalent of 600 tanker trucks with a 15,000-litre capacity each and represented daily profits of $200 million MXN ($10,297,800 USD) (ibid.). The losses of $7.5 billion USD over these three years represent most of the resources required for the government's planned refinery ($8 billion USD) (Rodríguez and Mendieta 2018).

The new government, though more transparent, has lacked consistency regarding data. In January 2019, the leader of Obrador's faction in congress declared that fuel trafficking represented losses of more than $80 billion MXN a year ($4.1 billion USD) (*Noticieros Televisa* 2019); a $10 billion MXN ($526 million USD) difference from the amount first given by the President. The first set of data presented by Obrador in late December 2018 did not match the numbers presented by PEMEX's Director in mid-January 2019 (presidential December data displayed pipeline theft only). This showed that the government had problems measuring the MFBM and that guesstimates are common when assessing criminal enterprises.

The $7.5 billion USD figure Obrador presented in December 2018 had a series of limitations. This amount was calculated considering 46,428,000 barrels of fuels were stolen between 2016 and 2018. In January 2019, the Director of PEMEX presented official data in which 54,385,000 barrels were stolen during that period (barrels taken from pipelines, refineries and distribution terminals). To this discrepancy we can add that stolen barrels reported for 2016 did not include pipeline losses and, considering the growth tendency and the fact that this fuel-theft source was omitted, we could presume that more than 64 million barrels were stolen between 2016, 2017 and 2018.

A document that puts into serious question the data presented by President Obrador is the PEMEX 2018 report to the US Securities and Exchange Commission. This report, which is the most comprehensive assessments on PEMEX's financial and operative situation, presents a scenario regarding the MFBM between 2016 and 2018 that differs substantially from the one presented by the President. The losses resulting from illicit actions amounted to $9.09 billion MXN in 2016 ($440 million USD), $22.9 billion MXN in 2017 ($1.16 billion USD) and $39.4 billion MXN in 2018 ($2 billion USD). In this three-year period, $3.6 billion USD worth of fuels were stolen from PEMEX according to this report (Fig. 3.4).

Even though these numbers differ significantly from Obrador's data on the MFBM, they point to a multibillion-dollar illicit enterprise showing a pressing growth. Between 2016 and 2017 non-operating losses due to illicit activities grew by more than 164% and by 72.6% between 2017 and 2018. According to the same report, in 2018 the average theft of fuel amounted to approximately 55,900 barrels a day ("Annual Report Pursuant..." 2018a, 183), the equivalent to 8,888,100 litres and an annual average of approximately 20,403,500 barrels. These numbers are closer to those presented by Obrador in 2018.

PEMEX admitted having limitations to accurately measure the financial impact of the MFBM. In the report to the US Securities and Exchange Commission, PEMEX stated that by late December 2018, the company had limited control over financial reporting. This shortcoming translated into an ineffective *"design and implementation of controls providing*

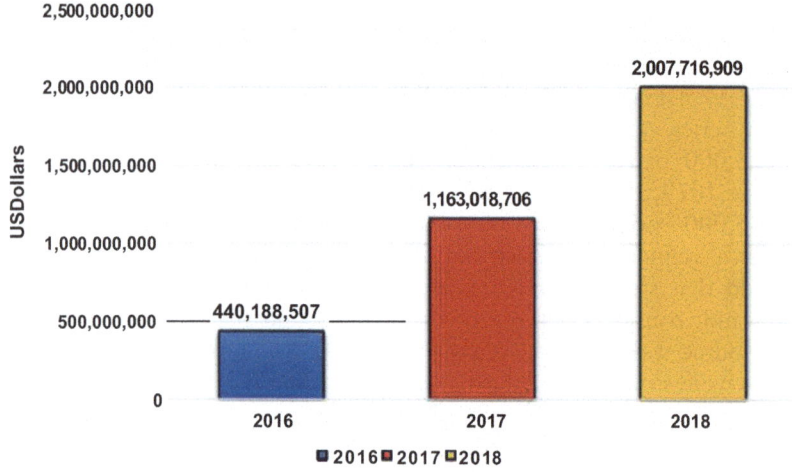

Fig. 3.4 PEMEX non-operating losses due to illicit activities ("Annual Report Pursuant..." 2018a, 319)

reasonable assurances regarding prevention of unauthorized disposition of assets" and this allowed the participation of *"certain employees involved in the illicit market in fuels"* ("Annual Report Pursuant..." 2018a, 183). The SOE goes further affirming that fuel trafficking increased partly because of ineffective controls and because the company *"did not have in place internal procedures to detect and investigate such matters"* (ibid.). This information reveals that PEMEX has problems measuring the MFBM accurately and that even the best data available will only give us an approximation of the size of this illicit market. These official figures, despite their inconsistencies, are the best option available to assess the MFBM.

The MFBM involves multiple criminal networks: from macro-networks to small criminal gangs with specialized mid-sized groups in between. They vary in their capacities and objectives: whether in organization level, territorial presence, weaponry, professional equipment and participation in other criminal enterprises. In these networks, coercive non-state actors work in coordination with law enforcement agents, current and former PEMEX employees, businessmen and communities. These groups include state and non-state actors uninvolved in the infamous armed wings of *sicarios*. Many of these networks not only exploit the MFBM, but also engage in other criminal activities like vehicle theft, extortion, homicides, kidnappings and money-laundering (EnergeA and Grupo Atalaya 2017, 18).

The MFBM requires control over territories to enable extraction. Extraction sources include refineries, distribution terminals and pipelines. The first two are usually found in urban centres while pipelines can be found in urban, semi-rural and rural areas. Fuel trafficking also entails controlling intermediary zones for storage and distribution (Fuerte Celis et al. 2018, 4). All the local fuel trafficking networks analysed for this research had actors specialized in transport and storage.

MFBM Operations: Installations and Pipelines

In January 2019, the presidency presented the following data on barrels stolen from PEMEX's installations (refineries and distribution terminals) and pipelines. The information presented by the federal government detailed how pipelines and installations have fluctuated in importance with regards to fuel trafficking (Fig. 3.5).

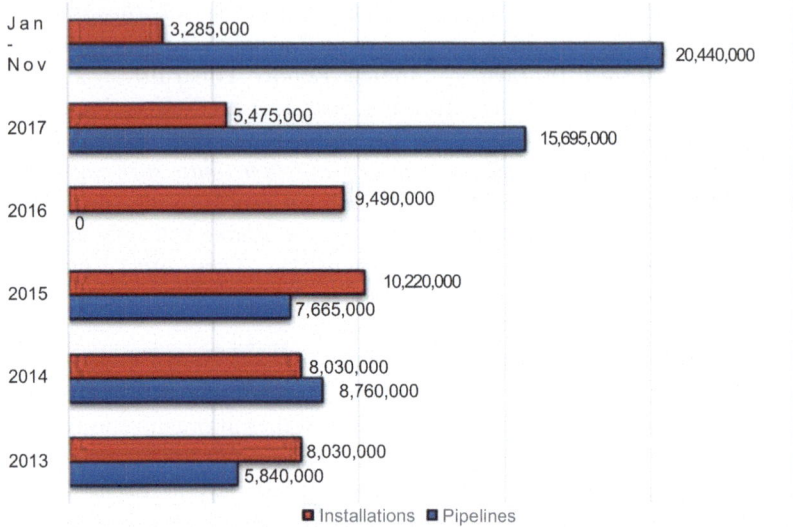

Fig. 3.5 Stolen barrels in PEMEX: installations vs pipelines (2013–2018) (Romero 2019a)

During this period, approximately 102,930,000 fuel barrels were stolen from PEMEX's refineries, terminals and pipelines. What's interesting is how these two sources of stolen fuels behave annually: from 2013 to 2015 fuel-theft operations were divided between installations and pipelines. This changed in 2017–2018, when pipelines became the predominant source. What these figures show is the versatility and the co-option criminal networks achieved within PEMEX, allowing to exploit the company in different areas swiftly changing their operations on a yearly basis (Table 3.1).

These changes show how the fluid structure of criminal networks makes them capable of rapid innovation (Bright and Delaney 2013, 239). When these groups find any trouble exploiting refineries, terminals or pipelines, they can adapt rapidly co-opting individuals with the required expertise and modify operations efficiently. Another concerning issue this data points to is the prominence of pipeline theft since 2017–2018, amounting to most of stolen fuels in that period. This is relevant because PEMEX transports between 60 and 70% of its refined products through pipelines (EnergeA and Grupo Atalaya 2017, 8) and that the network

Table 3.1 Stolen PEMEX barrels due to fuel-theft 2013–2018

Year	Barrels stolen	% stolen in installations	% stolen in pipelines
2013	13,870,000	58	42
2014	16,790,000	48	52
2015	17,885,000	57	43
2016	9,490,000	–	–
2017	21,170,000	26	74
2018	23,725,000	14	86

Romero (2019a)

covers 14,000 kilometres in size (ibid., 35–36). 45,968 illegal extraction points were found between 2011 and October 2018. This information reflects the magnitude of the challenge faced when monitoring such a colossal space facing adaptable criminals.

Next, fuel-theft modalities will be clarified. Consequently, the role of PEMEX grey actors will inevitably have to mentioned, given the fact that they enable the day-to-day operations of fuel trafficking groups. Their larger role will be developed in Chapter three, but here we will explore the functions they fulfil in fuel-theft operations.

Theft in Refineries and Storage and Distribution Terminals

Theft in refineries and terminals operates as follows: for a tanker truck to obtain fuel acquired by a buyer, the driver requires an invoice emitted by a retailer. This document is shown to the terminal or refinery porter and to the pump operator for the hydrocarbons to be delivered to the truck. By reusing one invoice or multiple forged invoices, drivers can load their tanker trucks on multiple occasions. In the process a tanker truck can load from 15,000 to 30,000 litres at a time (Pérez 2018a). In this process the actors involved can be seen in the following chart (all unionized except for the Terminal Manager, a middle-management non-unionized employee) (Illustration 3.2).

Clandestine pipelines installed in refineries have also been detected. In January 2019, the Army found in the Salamanca refinery a buried 3-kilometre pipeline connected to an external storage facility (López Obrador 2019a). In these installations production figures are altered to camouflage fuel-theft. Worryingly, PEMEX's Director admitted that

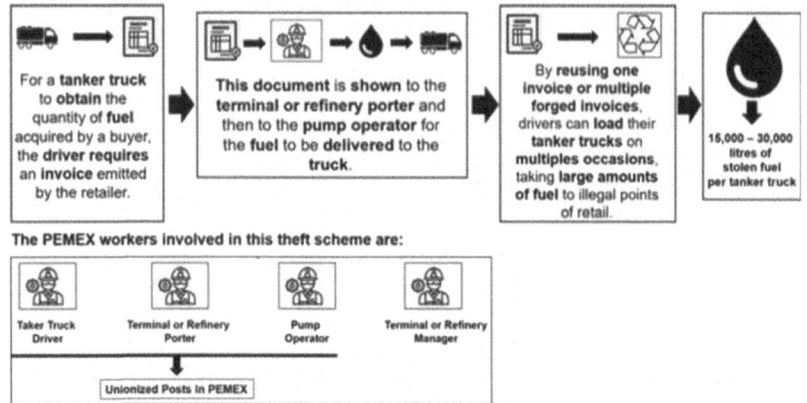

Illustration 3.2 Fuel-theft in refineries and distribution terminals (Elaborated by author with information from Pérez 2018a)

authorities are still placing control mechanisms to assure that no storage and distribution terminals or refineries have "*differences in production*" (Romero 2019b). This lack of control systems could partly explain the guesstimate tendency of the government.

Pipeline Theft

Between 2011 and 2015 approximately 50,245,509 fuel barrels were stolen from PEMEX pipelines. Of this total, 59.5% took place in a pipeline known as a polyduct, which transports different refined products (Table 3.2).

Mexico's polyduct network is divided into six systems. Out of all these, the *Sur-Centro-Golfo-Occidente* is the largest (8662 kilometres) and the one with most volumetric losses and illegal syphons between 2011 and 2015 (Table 3.3).

Pipeline extraction accounted for 67,890,000 stolen barrels between 2013 and November 2018 (López Obrador 2019b). During the Fox administration (2000–2006) one illegal syphon was detected daily, increasing to 7.6 between 2009 and 2016 (EnergeA and Grupo Atalaya 2017, 35). Fuel trafficking's explosive growth in 2011, especially in the

Table 3.2 Fuel-theft in PEMEX pipelines by fuel type 2011–2015

Fuel type	Total losses in barrels	Illegal extraction points
Magna gasoline	−435,794	7
Fuel–oil	−145,083	4
Diesel	−457,122	20
Oil	−19,294,330	1253
Jet fuel	−11,484	4
Polyduct	−29,897,351	14,554
Premium gasoline	−4345	5

EnergeA and Grupo Atalaya (2017, 31–32)

Table 3.3 Ranking of polyduct systems by volumetric losses and illegal extraction points 2011–2015

Position	System	Volumetric losses in barrels	Illegal extraction points
1.	Sur Centro Golfo Occidente	−16,021,925	8836
2.	Norte	−11,412,121	3320
3.	Topolobampo	−2,194,209	1615
4.	Rosarito	−170,723	454
5.	Guaymas	−98,373	168
6.	Progreso	0	0

EnergeA and Grupo Atalaya (2017, 33)

Sur-Centro-Golfo-Occidente system, reconfigured the geography of criminality in Mexico, displacing it towards the centre regions of the country (Map 3.2).

This theft modality requires the involvement of PEMEX grey actors including the pipeline management unit, its head manager, operations chief, maintenance chief, line engineers, supervisors, pumping station chiefs and instrumentations and control supervisors. All these posts, except the manager, are unionized. The subsidiary *Pemex Logística* also has areas that are involved in the management and operation of pipelines and have had a role in fuel-theft (Illustration 3.3).

Other important actors are unionized workers with the expertise and equipment for pipeline tapping. They can instal "professional" removable extraction valves for $50,000–$150,000 MXN ($2600–$7800 USD). A

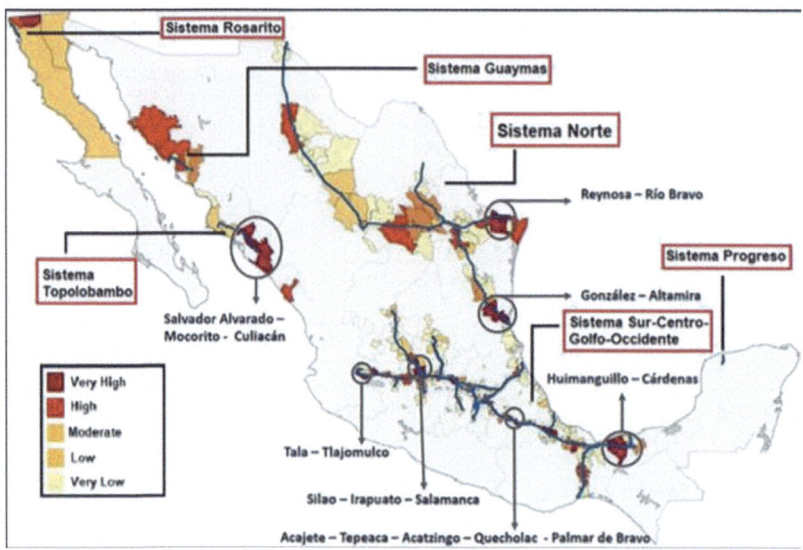

Map 3.2 Polyduct systems and regions with high concentration of illegal extraction points 2009–2016 (EnergeA and Grupo Atalaya 2017, 38)

pipeline theft operation lasts from 50 minutes to 2 hours and uses tanker trucks with a carrying capacity of 20,000–43,000 litres (Osorno 2019). The involvement of PEMEX grey actors in pipeline fuel-theft has been acknowledged by federal authorities. In October 2019, the Director of PEMEX stated in a congress hearing that in a case that took place on the 18th and 19th of December 2019 in the Tuxpan–Azcapotzalco pipeline "*there was a pressure drop of 4,000 barrels to 1,000*" (Romero 2019b) that lasted for 12 hours. This drop remained unreported despite detection by the monitoring unit and the fact that company protocols establish that, when a pressure drop equivalent to 200 barrels is recorded for 2 hours, it is necessary to close the pipeline. "*The pressure was falling and instead of closing the pipeline it was kept open, knowing that something was happening*" (ibid.) sentenced the Director.

To the co-option of grey actors, we can add the challenge posed by the technical limitations of pipeline security systems. There is the conception that pipeline monitoring systems can detect exactly where extraction incidents take place, though the reality paints a more intricate picture.

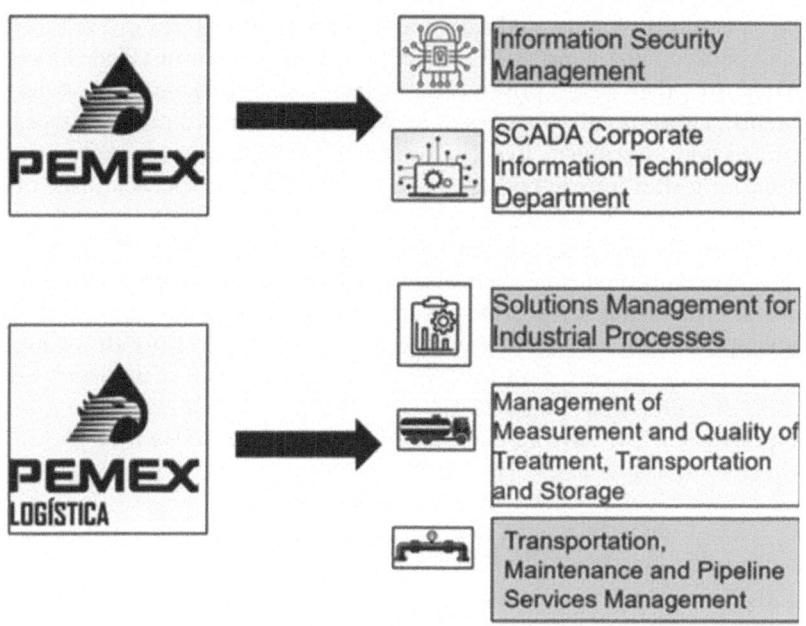

Illustration 3.3 PEMEX pipeline management areas (Elaborated by author with information of Flores 2019)

The main monitoring pipeline system used by PEMEX is the Supervisory Control and Data Acquisition system (SCADA) that permits the overseeing of *"the operating conditions of the pipelines, such as pressure, flow, temperature and packing, and above all, it allows to remotely close 505 of the 1,302 valves that exist along the pipeline system to stop fuels from flowing at the points where an irregularity is detected"* (EnergeA and Grupo Atalaya 2017, 60). By closing the flow of fuels between valves, SCADA can diminish theft losses, even though any derivatives left in a section between closed valves will still be available for extraction.

The SCADA is a system that allows to reduce fuel losses to an extent, but it has limitations. For example, it only allows the closure of under 38% of the valves in PEMEX's pipeline network. To this we can add that SCADA has also fallen victim to the state's corruption: in February 2019 it was revealed that 170 of 379 monitoring sites of the system (45%) were non-operational due to vandalism and lack of repairs (40 sites were sabotaged between 2015 and 2018) while 98 sites were never installed (Flores 2019). In other cases, private contractors installed incomplete useless systems. These monitoring sites of SCADA fall under two category types: manned and unmanned. Even in key energy installations where manned monitoring sites were supposed to be installed, this never took place (as in the Salamanca Refinery).

To be effective, SCADA must be complemented with a capacity to physically tend to pipeline theft that both PEMEX and the Mexican state have shown unable to provide. Moreover, SCADA is not as precise as public perception suggests. The system only reports pressure drops in a determined range between valves, where in *"some sections of the network it means distances of more than 15 kilometres [...], and if the extraction point is "professional", it may not be detected no matter the amount of physical inspections that are made"* (EnergeA and Grupo Atalaya 2017, 61).

Maritime Fuel Trafficking

PEMEX operates 15 maritime facilities in Mexico. These installations are complemented by 6 refineries and 77 storage and distribution terminals (Map 3.3).

Marine PEMEX operations have experienced criminal intervention beyond fuel trafficking, showing once more the diversification trend amongst criminal networks. According to the Navy Secretariat (SEMAR), criminal networks in the Gulf of Mexico involved in fuel trafficking target oil platforms (Pérez 2018b). Well-organized groups have stolen sophisticated equipment from platforms on several occasions (Guerrero Claudia 2018a, 2018b). Criminal attacks on oil platforms deserve further attention. PEMEX reported losing at least $224 million MXN ($11,353,743 USD) due to maritime criminal attacks on platforms off the coast of Campeche and Tabasco between 2016 and 2018 (Transparency Response 1857500102018).

Fishermen who saw these groups described them as armed commandos with highspeed boats (Reza 2019), stealing valves, batteries, cables, lighting installations, aluminium beams, monitors, drilling equipment, firewater hydrants and other types of valuable equipment. One incident reported by PEMEX stands out: on May 25, 2018, 40% of the heliport in the platform *Tsimin-B* was stolen with a value of $20 million MXN ($1,002,345 USD) (Transparency Response 1857500102018). This is

Map 3.3 PEMEX strategic energy installations (EnergeA and Grupo Atalaya 2017, 9)

one of multiple incidents where the values of stolen equipment go beyond $10 million MXN ($550,103 USD). In 2016, 48 of these incidents took place and by 2018 they increased to 197.

Equipment is not the only target of criminal groups operating in the oil-rich Gulf of Mexico. On January 15, 2019, the presidency made public that criminal actors were stealing crude oil from platforms in the maritime zone *Sonda Campeche* (López 2019). According to dissident union leader José Luis Rivero de la Rosa, the tankers used in this area have a carrying capacity of 80,000–100,000 barrels. Considering that, during 2018, the Mexican barrel's price fluctuated between $45 and $77 USD (Banco de México 2020), profits from a single vessel are substantial. For these operations to take place, vessels are connected to anchored offshore buoys that load or unload hydrocarbons. It is believed the crude oil is then taken into open waters for ship-to-ship transfers that are destined for the US or South America (López 2019). Illegal ship-to-ship fuel transfers have been documented in Southeast Asia, West Africa and the Mediterranean (Ralby and Soud 2018).

The government has expressed outrage at private shipping companies for their involvement in maritime fuel trafficking in the coasts of Campeche and Tabasco. Navy Secretary Rafael Ojeda mentioned that, in incidents where their forces caught vessels trafficking hydrocarbons, the companies involved took no actions against captains and crews. He also stated that piracy operations against oil platforms and private ships take only around three to four minutes: "*They already know what they are going to take, and many times what they are going to steal is on deck, there is collusion*" (as cited in Guerrero Claudia 2020b). Grey actors are crucial to criminal operations even in the waters of the Gulf of Mexico.

Oil platform piracy incidents have taken place in parallel to increased attacks on ships in the Gulf of Mexico. Since 2017 around a hundred piracy ship attacks have taken place off the coasts of Campeche and Tabasco (*El Universal* 2019). SEMAR has cited the participation of large criminal networks like the Zetas as well as smaller specialized groups. Oil platforms and marine vessel theft cases are not the only instances where criminal networks have shown interest in PEMEX. In 2018, it was revealed during the trial of Joaquín Guzmán that the capo held a meeting with PEMEX officials in 2007 to discuss the possibility of using the SOE's tankers to transport cocaine (Feuer 2018).

Criminal networks use PEMEX oil tankers, shipping companies and fishing vessels for maritime fuel trafficking. In 2018 SEMAR investigated 90 vessels suspected of being involved in fuel trafficking using forged invoices (Pérez 2018b). These operations largely supply internal demand (ibid., Pérez 2018a), but there are also signs of transnational smuggling. In relation to the export side of the MFBM, cross-checking sources proved relevant. One mention of the exporting side of the MFBM came from SEMAR, which was reinforced when cross-checked with information of the CRE. This source pointed to two regions in Mexico's polyduct systems that are linked to transnational fuel trafficking: In Reynosa-Río Bravo, Tamaulipas criminal networks "clone" "*companies and tanker trucks that distribute crude oil and natural gas to refineries and fuel retailers*" in Texas, making it an "*area that has registered an extensive illegal logistics network that ensures the extraction of hydrocarbons in pipeline in Mexico for its later storage*" in American tanker trucks (EnergeA and Grupo Atalaya 2017, 108). The Huimanguillo-Cárdenas region in Tabasco exhibits fuel trafficking to Central and South America using tankers for marine transport and camouflaging operations through legal and façade companies (ibid.).

There are documented cases of transborder smuggling into Mexico since the early 2000s. Information of the Mexican senate presented in 2002 warned that, because of PEMEX being unable to maintain competitive fuel prices, derivative smuggling coming into Mexico from both the north and southern borders was increasing (Amaya 2002, 117). According to this source, Pale Oil was being smuggled into Mexico and being used as a substitute for agricultural diesel. Traffickers sold this derivative in Mexico for $3.60 MXN/litre ($0.35 USD), while the diesel produced by PEMEX was $4.63 MXN/litre ($0.45 USD) (ibid.). Recently, transborder trafficking from Central America has been verified. In Chiapas, Guatemalan gasoline is sold at $3.78 MXN/litre ($0.2 USD) cheaper than the market price. One of the vendors said that, with the protection of community involvement, *"the business has always existed, although now more"* (Morales 2019).

Trafficking networks in the Gulf of Mexico are conformed by criminals, PEMEX workers, shipowners and crews. Other actors manage storage, commercialization and money-laundering. In this criminal puzzle, the refinery in Ciudad Madero, Tamaulipas plays a pivotal role. This refinery, the first in Latin America, occupies 544 hectares with 22 processing plants, storage facilities and a maritime terminal where gasolines, jet-fuel, diesel and other derivatives are manufactured. According to SEMAR, vessels owned by PEMEX coming from Campeche take their fuel cargo to the Ciudad Madero Refinery. Once there, trafficking vessels enter the refinery at night to extract fuel from the ships. The volumes of stolen fuel in maritime trafficking are significant: 600,000–800,000 litres per shipment (30–40 tanker trucks) (Pérez 2018b). This is one example of a broader trafficking phenomenon occurring all over Mexico's Gulf coast, where PEMEX operates seven marine facilities and with smuggling operations between Campeche, Tabasco, Veracruz and Tamaulipas.

Imported gasolines enter Mexico through the marine terminals of Pajaritos, Rosarito, Topolobampo, Manzanillo, La Paz, Mazatlán and Tuxpan; with the last terminal concentrating 30% of total imports. The federal government announced in January 2020 that it was investigating these installations due to the theft of imported gasolines using tanker trucks (Guerrero Claudia 2020a).

Nautical trafficking is arduous, an operation can take 6–8 days of sea travel depending on weather conditions, increasing the risk of detection. As a response, official invoices are obtained to camouflage marine trafficking. This is obtained by co-opting PEMEX's accountants, engineers and refineries and terminals personnel. Workers of the SOE act as facilitators of logistical arrangements and provide access to documentation which allows state and non-state actors to appear to be "lawfully" transporting or commercializing hydrocarbons of illegal origin (Farah 2013, 78). PEMEX workers, for example, communicate to marine traffickers to access the company's facilities.

All the modalities of theft offer diverse opportunities for criminal networks. And this is just one of many other incentives offered by the MFBM.

The Incentives of the MFBM

In this section the incentives that make the MFBM an appealing criminal enterprise will be developed. These include low costs, impunity, multiple distribution outlets, camouflaging options and the possibilities of occupying the legal economy.

Former intelligence official Gustavo Mohar revealed that a PEMEX official in Hidalgo admitted that in a local pipeline, a criminal network stole between $10 and $15 million MXN ($519,861–$779,791 USD) of fuels monthly by syphoning for two-hour periods. Mohar estimates it costs $3000 MXN ($155 USD) to instal an illegal pipeline tapping. These figures reveal the profits criminal networks access through fuel trafficking with negligible operating costs. Additionally, the MFBM became more profitable due to abrupt price increases occurring after 2010 (Mohar 2018). This information was corroborated by the statement of a former Zetas member, affirming that it was "*safer to grab money from where the pipelines were, since it was better paid and you had more time be free on the street*" (Osorno 2019).

Criminal networks access vast revenues via fuel-theft while the probability of being punished for it is minimal. Between 2001 and 2011, PEMEX presented 2611 legal complaints related to fuel-theft, only 15 concluded in a sentence (Garay-Salamanca and Salcedo-Albarán 2016, 118). This situation persisted: between 2015 and 2017 17,217 individuals were detained for fuel trafficking out of which 343 received a sentence (Muedano 2018). In this regard, the argument that insurgent groups can increase recruitment under certain conditions applies to this criminal market's expansion: if costs are low and benefits high, more individuals will get involved in the black market (Ross 2015, 115). This combination of profits and widespread impunity created a favourable context for criminals to increase their participation in the MFBM.

But impunity is one of more incentives. Once fuels are stolen the options for distribution are diverse, this also makes the MFBM appealing. Stolen fuels can be sold at wholesale (cement producers and gas stations) and retail levels (stations, sales in highways) (EnergeA and Grupo Atalaya 2017, 5). The MFBM gives criminals the possibility to access profits in less time. For example, while *"cocaine takes an average of six months to produce"* (Park 2016, 28), stealing hydrocarbons gives criminal networks access to valuable goods without prolonged production times or the necessity of building laboratories or producing crops. Methamphetamine production requires access to sophisticated export structures because precursors are obtained in India (Reuter 2014, 5). The MFBM concurs with the trend observed in Latin America where regional black markets have thrived occupying high and low profit enterprises generating substantial revenues (Yashar 2018, 72). Fuel trafficking is highly profitable, time-efficient and a less complicated enterprise.

The MFBM enables the possibility of selling goods through legitimate retailers. Pérez (2018a) and Mohar (2018) documented cases of gas station owners being coerced into selling stolen fuels. In 2011 when the Organization of Petroleum Retailers filed a criminal complaint alleging criminal groups were forcing gas station owners to sell stolen fuels, stressing the numerous cases in Coahuila, Michoacán, Nuevo León and Veracruz (Knott 2011). Likewise, hydrocarbons allow criminals to camouflage their operations by forging documentation, equipment and transport. This works as an additional protection barrier against detection. Some examples include:

1. A truck forged as a military vehicle was detained transporting 200 litres of diesel. The passengers were carrying PEMEX IDs (Milenio Digital 2017).
2. Gas station owners denounced men approaching them with forged commercialization permits from Mexico's energy regulator, offering fuels in bulk at lower prices than the legal market (Meana 2017).
3. In 2011 a former soldier witnessed how the Zetas cloned uniforms and trucks from SEMAR to transport stolen fuel in Veracruz (Harp 2018).

4. There are cases in which criminal networks instal gas stations to sell stolen fuels. This was first mentioned to me by a risk consultant (Salazar 2018). In 2011, an investigation was launched on a gas station operating in Zapopan, Jalisco. According to PEMEX, this station was installed using forged documentation and appeared in an official database as a station located in Tamaulipas (Franco 2011).

These examples show the lengths that criminal networks will go to exploit the MFBM, developing multiple camouflaging methods.

Criminal networks have used companies to profit from PEMEX as contractors, further increasing their incentives to participate in the MFBM and develop energy companies. One example is the Zetas, who breached PEMEX through the company *ADT Petroservicios* giving them access to vast revenues. This company became a PEMEX contractor for building projects and providing drilling and cleaning services. Between 2003 and 2011 *ADT Petroservicios* formalized thirty contracts with PEMEX for $2 billion MXN ($170 million USD). This has led to comparing the Zetas with security/energy services conglomerate Halliburton (Correa-Cabrera 2017) and exemplifies how legality and illegality operate jointly.

The Actors in the MFBM

Having explained the diversification of Mexican criminal groups, we must next define these entities. Criminal groups, far from being hierarchical, are "*fluid social systems where flexible exchange networks expand and retract according to market opportunities and regulatory constraints*" (Kenney 2007, 234). Research on criminality has agreed that what was once labelled as "organized crime" are mercurial and adaptable network structures coordinating different illicit enterprises. A criminal network is a succession of linked nodes that can represent individuals or organizations all connecting to obtain unlawful revenues.

Criminal networks are largely driven by market forces and the best way to understand them is as supply chains that move products between suppliers and consumers. Illicit networks are composed of individuals, organizations, processes, technological inputs, etc. Members maintain roles within the network to "*enable key processes and functions to operate without interference from law enforcement*" (Deville 2013, 65). As long as essential roles are preserved, the network will adapt to disruption to sustain the supply chain.

Criminal networks can come into existence through individuals who deliberately exploit an illegal activity. Another option is a network that builds itself through market inertia. Williams (2001) names these "*directed networks*" and "*transaction networks*", respectively. Kenney (2007) has a similar typology: "*wheel networks*" organized and managed by a core enterprise, and decentralized and self-organizing "*chain networks*", operating with autonomous nodes performing tasks with other nodes without oversight (243–245). Colombian traffickers of the 1980s and 1990s were directed networks, organized since their inception to smuggle cocaine into the US. By contrast, heroin trade in Southeast Asia is dominated by transaction networks in which "*brokers play a critical role at almost every stage of the process. Producers supply heroin to independent distributors, and it is then passed along a chain of brokers until it reaches the retail market*" (Williams 2001, 69). These networks are not mutually exclusive: a directed network could be part of a transactional network.

Academic consensus about the relevance of networks has been recognized by law enforcement. For example, German Federal Police "*has observed that most of the criminal organizations it investigates are 'loose, temporary networks'*" while "*hierarchical structures are*" the minority (ibid., 63). This understanding has gained recognition, but exceptions persist. In Mexico what are still defined as "drug trafficking cartels" are groups without a vertical structure and diversified beyond narcotics. These groups vary in size, some can be local while others regional and transnational. This concurs with the description given by Williams regarding the scope of criminal networks, ranging "*from small, very limited associations at the local level to transnational supplier networks*" (ibid., 69).

Networks adapt by making alliances, conflicts, infiltrations, centralizations or decentralizations of their structures given the circumstances. These structures lack centralized leaderships, which makes them resilient to law enforcement. They operate without large infrastructure investments making them mobile. Zaitch and Antonopoulos (2019) argue, after analysing literature on criminal networks in Latin America of the last 15 years, that these groups have shown remarkable capability to learn, innovate and adapt "*in order to stay in business*". This adaptability allowed criminal networks to expand their enterprises "*to regions and countries far*" from their territories of origin, enabling alliances with other criminal groups and sustaining "*a diversification of illegal activities*" (2–3).

Their structures separate members into loosely organized specialized cells, while core leaderships use brokers to protect themselves from direct intervention (Kenney 2007, 241). Participants only understand their function without knowing key players or crucial information that could threaten the network's existence. It has been confirmed, for example, that this is the organizational model of Colombian criminal networks (ibid., 246). Additionally, criminal networks show great capacity to create redundancy amongst nodes, largely amongst the ones belonging to specialized peripheral positions (Williams 2001, 80). This allows to dispose of compromised individuals or groups with minimum disruption. Operational knowledge throughout nodes is compartmentalized, while redundant nodes hold "*simplified roles*" (Deville 2013, 65) to limit disturbance in the overall operations of the network. These features make networks resilient, permitting them "*to maintain organizational integrity even in an extremely inhospitable environment*" (Williams 2001, 81).

Not only are these networks built for resilience and adaptation, but their operational capacity is further strengthened by conventional anti-crime strategies that rely on conceiving these groups as "*hierarchical organizations with a strong top-down structure*" (Farah 2013, 91) and prioritize targeting criminal leaders. This allows multiple grey actors (in financial, political and logistical roles) to operate without scrutiny, allowing for network continuity despite crackdowns. Eliminating network participants leads to adjustments and continuity, as observed in as diverse cases as Al-Qaeda, Colombian traffickers and Canadian drug importers (Carrington 2016, 49).

When necessary, criminal networks modify their structure "*by expanding to gain access to necessary resources, skills or knowledge*" (Bright and Delaney 2013, 239). They can adapt rapidly by integrating individuals with the required expertise, increasing their scope and diversifying their ranks with "*speed and low cost*" (ibid.). The example of the changes in the exploitation of PEMEX installations attest to this. To exploit pipelines and installations, and alternate between them, these groups can dispose of certain actors and integrate others efficiently, making them "*fluid and temporal*" (Mudde 2019, 50). And it is in this process that grey actors become crucial.

Groups engaged in sophisticated criminal enterprises constantly require economic, political, institutional, technological and legal resources. In the case of the MFBM, some of these resources can include guaranteeing impunity and enabling network-expanding opportunities, resources that can be provided by high-level officials. Williams (2001) argues these agents are *"network extenders"* (81). Other resources include technical inputs from PEMEX workers or entrepreneurs (pipeline extraction, installation access, money-laundering). Without these resources, a criminal market like the MFBM could not thrive (Garay-Salamanca and Salcedo-Albarán 2016, 46), and grey actors provide them through their presence in the legal and illegal realms. If a criminal organization has more grey actors, resource access will be more diversified and will make the group resourceful and resilient.

Garay-Salamanca and Salcedo-Albarán (2016) define criminal networks as a set of agents (legal, illegal and grey) collaborating to exchange financial, political and social resources of individuals, companies and entities (48–49). The diversification towards different illegal activities supports the understanding of criminal groups not as cartels but as networks integrated by state and non-state actors. *Macro-criminal networks* (ibid.) are conformed by agents varying between hundreds operating across multiple jurisdictions. Macro-criminal networks can develop sub-networks, specialized in different illegal enterprises.

The MFBM involves the participation of criminal networks of different capacities, including large criminal networks and specialized sub-networks. The methods implemented by Mexican criminal networks to traffic derivatives continuously evolve in response to government measures to curb their activities. To establish a more nuanced understanding of the actors involved the following typology is proposed (Illustration 3.4).

Large Criminal Networks	- The FBM has become an important illegal activity for these groups. - They operate in multiple states and regions across the country. - They have sophisticated technical knowledge, specialized tools and weapons (for example, they use pressure valves for pipeline extraction that prevent fuel spills). - They have the capacity to "clone" official documentation, uniforms and tank trucks. - They use different strategies to exploit this market (e.g. they threaten municipal authorities to cover up their operations and force gas station owners to sell and distribute stolen fuel). - They have the capacity of co-opting local crime groups and create specialized sub-networks. - Two examples of such groups are the CJNG and the Zetas.
Specialized Fuel Trafficking Groups	- Their operation capacity is at the state level (when they have multiple state presence it is usually in neighbouring areas) and various municipalities. - They are mid-sized when compared to their larger counterparts. - They have technical knowledge and equipment to make clandestine hermetic extraction points. - They have armed cells specialized in various activities such as: surveillance, drilling, custody and transport of tank trucks. - These groups can confront rival gangs or groups specialized in stealing fuel for large criminal networks. - Their main source of income is fuel theft and trafficking. - One example of these types of networks is the Santa Rosa de Lima Cartel.
Fuel Criminal Gangs	- They operate at the municipal level. - They are groups with more rudimentary operating systems. - They drill pipelines with tools such as saws, hammers and chisels. - Their operations are more linked to accidents such as spills and fires. - They transport the stolen fuel in smaller scale in plastic containers, using vans. - They tend to sell at a retail level, especially in highways. - Groups such as these operate in Puebla and Guanajuato.

Illustration 3.4 MFBM criminal network typology (Elaborated by author with information from EnergeA and Grupo Atalaya 2017, 19–21)

Bibliography

"Clonan camión de la SEDENA para robar combustible." *Milenio Digital*, April 17, 2017. http://www.milenio.com/estados/clonan-camion-de-la-sedena-para-robar-combustible. Date accessed: July 11, 2021.

"Desabasto de gasolina y cierre de ductos, análisis en Despierta." *Noticieros Televisa*, January 8, 2019. https://noticieros.televisa.com/ultimas-noticias/desabasto-gasolina-cierre-ductos-analisis-despierta/. Date accessed: July 10, 2021.
"Memoria de Labores 2007." Petróleos Mexicanos, 7–151.
"Memoria de Labores 2009." Petróleos Mexicanos, 7–207.
"Memoria de Labores 2011." Petróleos Mexicanos, 7–193.
"Mexican Energy Reform: Opportunity Knocks." *Deloitte*, 2014.
"Pemex perdió más de 17.000 millones de dólares en 2017, un 75% más que el año anterior." *El País*, February 27, 2018. https://elpais.com/economia/2018/02/27/actualidad/1519750558_185841.html. Date accessed: July 10, 2021.
"Pemex se dobla ante el chantaje de los Zetas." *Proceso*, June 18, 2010. https://www.proceso.com.mx/nacional/2010/6/18/pemex-se-dobla-ante-el-chantaje-de-los-zetas-6788.html. Date accessed: July 10, 2021.
"Piratas atacan barco italiano en el Golfo de México." *El Universal*, November 12, 2019. https://www.eluniversal.com.mx/estados/piratas-atacan-barco-italiano-en-el-golfo-de-mexico. Date accessed: July 11, 2021.
Aguilar, Ruben. 2018. "125,000 muertos y la posibilidad de superar esa cantidad." *El Economista*, November 22. https://www.eleconomista.com.mx/opinion/125000-muertos-y-la-posibilidad-de-superar-esa-cantidad-20181122-0176.html. Date accessed: July 10, 2021.
Amaya, Téllez Rodimiro. 2002. *Legislando por Baja California Sur*, 117–118. México: Senado de la República.
Atuesta, Laura, and Yocelyn Pérez-Dávila. 2018. "Fragmentation and Cooperation: The Evolution of Organized Crime in Mexico." *Trends in Organized Crime* 21: 235–261.
Banco de México. 2020. "Precio de la Mezcla Mexicana de Petróleo." Accessed on February 14, 2020. https://www.banxico.org.mx/apps/gc/precios-spot-del-petroleo-gra.html. Date accessed: July 11, 2021.
Beckert, Jens, and Matías Dewey. 2017. "The Social Organization of Illegal Markets." In *The Architecture of Illegal Markets: Towards an Economic Sociology of Illegality in the Economy*, edited by Jens Beckert and Matías Dewey, 1–38. Oxford University Press.
Bright, David A., and Jordan J. Delaney. 2013. "Evolution of a Drug Trafficking Network: Mapping Changes in Network Structure and Function Across Time." *Global Crime* 14 (2–3): 238–260.
Carranza, Fabián. 2018. "La magnitud de las tomas clandestinas de hidrocarburos." *Transferencia TEC*, December 3. https://transferencia.tec.mx/2018/12/03/la-magnitud-de-las-tomas-clandestinas-de-hidrocarburos/. Date accessed: July 10, 2021.

Carrington, Peter J. 2016. "Crime and Social Network Analysis." In *The SAGE Handbook of Social Network Analysis*, 236–255. Sage.
Cervantes, Jesusa. 2012. "Las Mafias Desangran a PEMEX." *Proceso*, March 18.
Córdoba, Mayela. 2003. "Provoca ordeña de ductos reparaciones millonarias." *Reforma*, April 17.
Correa-Cabrera, Guadalupe. 2017. *Los Zetas Inc.: Criminal Corporations, Energy, and Civil War in Mexico*, 1–341. University of Texas Press.
Deville, Duncan. 2013. "The Illicit Supply Chain." In *Convergence: Illicit Networks and National Security in the Age of Globalization*, edited by Michael Miklaucic and Jacqueline Brewer, 63–74. National Defense University Press.
Durán-Martínez, Angélica. 2018. *The Politics of Drug Violence: Criminals, Cops and Politicians in Colombia and Mexico*, 1–299. Oxford University Press.
EnergeA and Grupo Atalaya. 2017. "Estudio para analizar la problemática de seguridad física en las instalaciones del sector hidrocarburos y emitir recomendaciones para el reconocimiento de costos por concepto de seguridad que la Comisión Reguladora de Energía lleva a cabo en sus procesos de revisión tarifaria," 1–368.
Farah, Douglas. 2013. "Fixers, Super Fixers, and Shadow Facilitators: How Networks Connect." In *Convergence: Illicit Networks and National Security in the Age of Globalization*, edited by Michael Miklaucic and Jacqueline Brewer, 75–95. National Defense University Press.
Felbab-Brown, Vanda. 2019. *Mexico's Out-of-Control Criminal Market*, 1–29. The Brookings Institution.
Feuer, Alan. 2018. "The El Chapo Trial: New York's Newest Tourist Destination." *New York Times*, December 18. https://www.nytimes.com/2018/12/18/nyregion/el-chapo-trial-tourism.html. Date accessed: July 11, 2021.
Fitch Solutions. 2019. "Mexico Oil & Gas Report Q1 2019," 5–107.
Flores, Nancy. 2019. "Inservibles 170 sitios estratégicos de monitoreo antihuachicol en PEMEX." *Contralínea*, February 18. https://contralinea.com.mx/inservibles-170-sitios-estrategicos-de-monitoreo-antihuachicol-en-pemex/. Date accessed: July 9, 2021.
Franco, Gilberto. 2011. "Investigan en Jalisco una gasolinera pirata." *El Norte*, June 23.
Fuerte Celis, María del Pilar, Enrique Pérez Lujan, and Rodrigo Cordova Ponce. 2018. "Organized Crime, Violence, and Territorial Dispute in Mexico (2007–2011)." *Trends in Organized Crime* 22: 188–209.
Garay-Salamanca, Luis Jorge, and Eduardo Salcedo-Albarán. 2016. *Macro-Criminalidad: Complejidad y Resiliencia de las Redes Criminales*, 1–191. iUniverse.
González, Rodarte Jorge. 2001. "El sindicalismo petrolero mexicano en perspectiva: 1911–1989." *Perspectivas Históricas* (January–June): 111–156.

Guerrero, Claudia. 2018a. "Llega crimen hasta altamar." *Reforma*, October 9. https://www.reforma.com/aplicaciones/articulo/default.aspx?id=1510717&v=4. Date accessed: July 11, 2021.

Guerrero, Claudia. 2018b. "Sospechan de obreros y contratistas." *Reforma*, October 10. https://www.reforma.com/aplicaciones/articulo/default.aspx?id=1511642&v=8. Date accessed: July 11, 2021.

Guerrero, Claudia. 2020a. "Van en puertos contra huachicol." *Reforma*, January 18.

Guerrero, Claudia. 2020b. "Acusa Semar omisiones de empresarios en robos." *Reforma*, April 24. https://www.reforma.com/acusa-semar-omisiones-de-empresarios-en-robos/ar1927483?v=4. Date accessed: July 11, 2021.

Harp, Seth. 2018. "Sangre y Petróleo Parte Dos." *Rolling Stone*, September 14.

Hernández, Alma, and Mayela Córdoba. 2009. "Cuesta ordeña $9 mil millones." *Reforma*, July 31.

Hope, Alejandro. 2017. "En Tiempos de Peña Nieto." *Nexos*, January 1. https://www.nexos.com.mx/?p=30852. Date accessed: July 10, 2021.

Instituto Nacional de Geografía y Estadística (INEGI). 2018. "Datos Preliminares Revelan que en 2017 se Registraron 31 mil 174 Homicidios."

Kenney, Michael. 2007. "The Architecture of Drug Trafficking: Network Forms of Organisation in the Colombian Cocaine Trade." *Global Crime* 3 (8) (August): 233–259.

Knott, Tracey. 2011. "Mexico Gas Vendors Forced to Buy Fuel Stolen by Gangs." *InSight Crime*, March 23. https://www.insightcrime.org/news/brief/mexico-gas-vendors-forced-to-buy-fuel-stolen-by-gangs/. Date accessed: July 11, 2021.

Lajous, Adrián. 2018. "Pemex en crisis." *Nexos*, June 7. https://www.nexos.com.mx/?p=37935. Date accessed: July 10, 2021.

Lessing, Benjamin. 2015. "Logics of Violence in Criminal War." *Journal of Conflict Resolution* 59 (8): 1486–1516.

López, Jorge X. 2019. "Se roban 75 MDD, a la semana, en crudo." *24 horas*, January 24. https://www.24-horas.mx/2019/01/24/se-roban-75-mdd-a-la-semana-en-crudo-infografia/. Date accessed: July 11, 2021.

López Obrador, Andrés Manuel. 2018. "Presidential Press Conference." Speech, Palacio Nacional, December 27.

López Obrador, Andrés Manuel. 2019a. "Presidential Press Conference." Presentation, Palacio Nacional, January 8.

López Obrador, Andrés Manuel. 2019b. "Presidential Press Conference." Presentation, Palacio Nacional, January 14.

Manuel, Llano and C. Flores. 2017. "Ductos, ¿por dónde circulan los hidrocarburos en México?" [map]. Scale 1: 3,500,000. México: CartoCrítica / Fundación Heinrich Böll.

Meana, Sergio. 2017. "Huachicoleros tienen permiso para vender gasolina, pero es clonado." *El Financiero*, May 17. http://www.elfinanciero.com.mx/economia/huachicoleros-tienen-permiso-para-vender-gasolina-pero-es-clonado. Date accessed: July 11, 2021.

Medellín, Jorge Alejandro. 2005. "Pemex: la vaca negra... 'ordeña' clandestina y robo." *El Universal*, March 21.

Menéndez, Eduardo. 2012. "Violencias en México: las explicaciones y las ausencias." *Alteridades* 22 (43): 177–192.

Miranda, Juan Carlos. 2018. "Pemex ahorró 11.2 mil mdp con despidos y el recorte de tiempos extra." *La Jornada*, March 4. https://www.jornada.com.mx/2018/03/04/economia/016n1eco. Date accessed: July 10, 2021.

Mohar, Gustavo (Risk consultant and former Intelligence Official). 2018. Interviewed by Samuel León in Mexico City, August 14.

Montalvo, Tania. 2017. "Pemex pierde 100 mil mdp por robo de combustible y fugas en el sexenio de Peña." *Animal Político*, February 2. https://www.animalpolitico.com/2017/02/robo-combustible-ductos-pemex-gobierno-pena/. Date accessed: July 10, 2021.

Morales, Mariana. 2019. "Trafican gasolina desde Guatemala." *Reforma*, May 21. https://www.reforma.com/trafican-gasolina-desde-guatemala/ar1681747. Date accessed: July 11, 2021.

Morales, Isidro. 2020. *The Future of Pemex: Return to the Rentier-State Model or Strengthen Energy Resiliency in Mexico?*, 1–45. Rice University's Baker Institute for Public Policy.

Morris, Stephen. 2013. "Drug Trafficking, Corruption, and Violence in Mexico: Mapping the Linkages." *Trends in Organized Crime* 16 (2): 207.

Mudde, Cas. 2019. *The Far Right Today*. 1st ed. Medford, MA: Polity Press.

Muedano, Marcos. 2018. "Fracasan sentencias contra huachicoleros." *Excélsior*, June 4. https://www.excelsior.com.mx/nacional/fracasan-sentencias-contra-huachicoleros/1242959. Date accessed: July 11, 2021.

Notimex. 2015. "Personal de Pemex descubre gasolina adulterada con etanol en Mérida, Yucatán." September 4. https://www.20minutos.com.mx/noticia/29049/0/pemex-descubre/gasolina-adulterada-etanol/merida-yucatan/. Date accessed: July 10, 2021.

Obregón, Óscar. 2016. "9 Puntos para Entender la Crisis de PEMEX." *Expansión*, April 13. https://expansion.mx/economia/2016/04/13/9-puntos-para-entender-la-crisis-de-pemex. Date accessed: July 10, 2021.

Osorno, Diego Enrique. 2019. "'Robar a Pemex es más redituable que ser zeta': Entrevista / Testimonio de un huachicolero." *Milenio*, January 18. http://www.milenio.com/policia/robar-a-pemex-es-mas-redituable-que-ser-zeta. Date accessed: July 11, 2021.

Oxford Business Group. 2018. "Reforms, Liberalisation Open Mexico's Energy Sector to Private Investment and Emphasise Clean, Sustainable Energy."

Consulted on November 28. https://oxfordbusinessgroup.com/overview/game-changer-sweeping-reforms-open-sector-private-investment-and-emphasise-clean-sustainable-energy. Date accessed: July 10, 2021.

Park, Jung H. 2016. "What Explains the Patterns of Diversification in Drug Trafficking Organizations?" Master's thesis, Naval Postgraduate School, Monterrey California, 1–85.

Pawley, Dawn. 2014. *Drug War Capitalism*, 4–212. AK Press.

Pérez, Ana Lilia. 2011. *El Cártel Negro: Cómo el crimen organizado se ha apoderado de Pemex*, 5–221. México: Grijalbo.

Pérez, Ana Lilia (Independent journalist and author). 2018a. Interviewed by Samuel León in Mexico City, August 23.

Pérez, Ana Lilia. 2018b. "Huachicoleros en el Golfo de México saquean Pemex con ayuda de empleados." *Animal Político*, October 22. https://www.animalpolitico.com/2018/10/huachicoleros-golfo-mexico-robo-combustible-pemex/. Date accessed: July 11, 2021.

Pérez, Ana Lilia. 2018c. "Huachicoleo a escala multimillonaria Dentro de Pemex, toda una industria paralela." *Proceso*, December 29.

Pérez-Treviño, Emma. 2008. "Valley Residents Among Dozens Arrested in Major Operation Targeting Gulf Cartel." *The Brownsville Herald*, September 8.

Petróleos Mexicanos. 2018a. "Annual Report Pursuant to Section 13 or 15(D) of the Securities Exchange Act of 1934," 3–432. United States Securities and Exchange Commission.

Petróleos Mexicanos. 2018b. "Informe Anual 2017," 4–141.

Ralby, Ian M., and David Soud. 2018. "Oil on the Water: Illicit Hydrocarbons Activity in the Maritime Domain," 2–17. Atlantic Council Global Energy Centre, April.

Raziel, Zedryk. 2018. "Alista AMLO plan contra huachicol." *Reforma*, December 7. https://www.reforma.com/aplicaciones/articulo/default.aspx?id=1558407&v=7. Date accessed: July 10, 2021.

Reforma. 2019. "Opera red de huachicoleo de NL a CDMX." Video. https://www.youtube.com/watch?v=UWQ00_XExQo. Date accessed: July 9, 2021.

Reinhart, Luke B. 2014. "The Aftermath of Mexico's Fuel-Theft Epidemic: Examining the Texas Black-Market and the Conspiracy to Trade in Stolen Condensate." *St. Mary's Law Journal* 45: 750–786.

Reuter, Peter. 2014. "Drug Markets and Organized Crime." In *The Oxford Handbook of Organized Crime*, edited by Letizia Paoli, 1–25. Oxford University Press.

Reza, Abraham. 2019. "Suben 310% ataques piratas a las plataformas de Pemex." *Milenio*, January 17.

Rodríguez, Silvia, and Susana Mendieta. 2018. "Nueva refinería, muy cara para el gobierno de AMLO: Moody's." *Milenio*, September

6. http://www.milenio.com/negocios/nueva-refineria-muy-cara-para-el-gob ierno-de-amlo-moody-s. Date accessed: July 10, 2021.
Romero, Octavio. 2019a. "Presidential Press Conferenece." Presentation. Palacio Nacional, January 14.
Romero, Octavio. 2019b. "Comparecencia del Ingeniero Octavio Romero Oropeza, Director General de Petróleos Mexicanos (PEMEX), ante las Comisiones Unidas de Energía e Infraestructura de la Cámara de Diputados LXIV Legislatura." Presentation. Cámara de Diputados, October 28.
Ross, Michael. 2015. "Conflict and Natural Resources: Is the Latin American and Caribbean Region Different from the Rest of the World?" In *Transparent Governance in an Age of Abundance: Experiences from the Extractive Industries in Latin America and the Caribbean*, edited by Juan Cruz Vierya and Malaika Masson, 109–136. Inter-American Development Bank
Rousseau, Isabelle. 2010. "PEMEX y la Política Petrolera: Los Retos Hacia el Futuro." In *Los grandes problemas de México*, edited by José Luis Méndez, 304–340. El Colegio de México.
Salazar, Rubén (Risk consultant). 2018. Interviewed by Samuel León in Mexico City, August 1.
Siglar, Edgar. 2018. "El Sexenio de Pemex: Crisis, Pérdidas y Recortes." *Expansión*, November 28. https://expansion.mx/empresas/2018/11/28/el-sexenio-de-pemex-crisis-perdidas-y-recortes. Date accessed: July 10, 2021.
Transparency Response "1857500102018." *PEMEX*, November 14, 2018.
Trejo, Guillermo, and Sandra Ley. 2016. "Federalismo, drogas y violencia: Por qué el conflicto partidista intergubernamental estimuló la violencia del narcotráfico en México." *Política y gobierno* 13 (1): 11–56.
Vicenteño, David. 2004a. "Toma la PFP más plantas de Pemex." *Reforma*, March 6.
Vicenteño, David. 2004b. "Alista PFP nueva fase de operativo en Pemex." *Reforma*, March 28.
Villamil, Genaro. 2018. "Nuevo aeropuerto del peñismo, las seis pistas de la corrupción." *Proceso*, March 27. Consulted on October 21. https://www.proceso.com.mx/opinion/2018/3/27/nuevo-aeropuerto-del-penismo-las-seis-pistas-de-la-corrupcion-202212.html. Date accessed: July 10, 2021.
Williams, Phil. 2001. "Transnational Criminal Networks." In *The Future of Terror, Crime, and Militancy*, edited by John Arquilla and David Ronfeldt, 61–97. RAND Corporation.
Yashar, Deborah J. 2018. *Homicidal Ecologies: Illicit Economies and Complicit States in Latin America*, 1–368. Cambridge University Press.
Zaitch, D., and G. A. Antonopoulos. 2019. "Organised Crime in Latin America: An Introduction to the Special Issue." *Trends in Organized Crime* 22 (March): 141–147.

PART III

A Journey from Monterrey to Mexico City

Numerous freight trucks await parked on the side of a two-lane highway. The sun is setting, painting the sky in shades of grey, blue and pink. The grand vastness is interrupted by isolated mountain formations, a staple of Mexico's northern landscapes. *"Look, as night falls... the business is booming"* says the voice of a truck driver to his co-pilot, who is recording the freight trucks as they drive down on the opposite side of the highway. The driver and his companion, a journalist, are travelling from Monterrey to Mexico City. Throughout Nuevo León, Coahuila, San Luis Potosí, Querétaro, Hidalgo, Guanajuato, Estado de México and Mexico City, thriving fuel trafficking operations are visible, as retailers offer their illegally obtained products on roads and highways.

During the day, camouflaged distribution is the common denominator: façade businesses like rundown tire shops, car repair shops, cafeterias, restaurants and stores are all used to sell stolen fuels while hiding in plain sight. When night falls, individuals with flashlights appear on the side of the roads to flash drivers, announcing the availability of stolen fuels. From inside the truck, they appear as shadows punctuated by bursts of light that capture all the attention.

The networks behind these operations offer 25-litre containers of fuels for anywhere between $350 and $380 MXN ($14–$15.2 MXN/litre) and upload the locations of their points of retail on Google Maps. Each point appearing on Google Maps has its own number, revealing a surprising level of organization given the rudimentary surroundings. During their journey, the journalist and driver make a stop in a small community in

San Luis Potosí, where most inhabitants make a living out of fuel trafficking. In this place, vans filled with containers are visible while freight transport trucks buy stolen fuels. Stained roads can be seen across the different highways and roads, due to fuel spills (Reforma 2019b). No matter the time of day, groups profiting from Mexican fuel trafficking operate without any major restrictions.

Reference

Reforma. 2019. "Opera red de huachicoleo de NL a CDMX". Video. https://www.youtube.com/watch?v=UWQ00_XExQo. Date Accessed July 9, 2021.

CHAPTER 4

Understanding the Growth of the Mexican Fuel Black Market

FACTORS BEHIND FUEL TRAFFICKING GROWTH IN MEXICO

This chapter will explore the factors behind the growth of the MFBM between 2011–2018. These factors are relevant but not totalizing, and others have played a role in the growth of this criminal market such as: the limitations of monitoring systems of the pipeline network, lacking resources of the PEMEX unit overseeing installation security or the absence of early efforts to curb this black market. These factors may not exhaust the complexity of the MFBM as a criminal phenomenon, but they are the most important ones, and their prioritization is the result of my fieldwork and the analysis of the insights shared by experts. Moreover, these factors, though applied to Mexico, can prove relevant to other contexts as they illustrate the connection between black markets, institutional fragility and economic transformations. Accordingly, these considerations led to the following factors: (1) fragmentation of criminal networks and black market diversification, (2) the co-option of grey actors and (3) fuel price increases in Mexico.

© The Author(s), under exclusive license to Springer Nature Switzerland AG 2025
S. León Sáez, *Mexico's Fuel Trafficking Phenomenon*, St Antony's Series,
https://doi.org/10.1007/978-3-031-70503-8_4

Fragmentation and Diversification of Criminal Networks

The causality of criminal fragmentation and illicit market diversification is not simple. The Zetas instituted a criminal diversification trend, critical in a context of widespread state crackdowns and enhanced criminal competition. Criminal groups have repeatedly shown that they can adapt practices among organizations (Ayling 2009, 192). In the Mexican case, the extended militarization of armed wings and aggressive market diversification became widely adopted strategies amongst different groups. Diversifying became necessary to ensure profits as drug trafficking became an increasingly challenging endeavour and smaller groups continuously formed, generating new conflicts alongside the ones with the Mexican state. *"As market-driven enterprises, Mexican illicit networks have a strong incentive to protect their profits. Violence is an effective means to that end"* (Deville 2013, 68). Black market revenues allow criminal networks to operate in opposition to the state and to dispute control over illicit supply chains with their criminal counterparts. Therefore, criminal fragmentation in recent years pushed the majority of black-market diversification.

Fuel trafficking growth in Mexico is a significant consequence of a criminal fragmentation and diversification phenomenon expanding in the country for more than a decade. Criminal diversification has been widely researched. Reuter (2014) traces its origins further back, stating that *"the expansion of the drug trade in the last 40 years has presented opportunities for pre-existing criminal groups to build on their core capacities in other activities"* such as gambling and prostitution (16). With a more contemporary take, Garzón (2014) explained that during the last decade, criminal groups have gone through a process of fragmentation and dispersion across Latin America. Multiple factions have replaced cartels and are involved in numerous criminal activities, particularly local ones, transitioning towards an organizational model known as *"predatory micro-networks"*.

When the federal government carried out its war on drugs in 2007, it became evident that criminal groups diversified their activities as narcotics became more dangerous (Fuerte Celis et al. 2018, 5). This diversification rested on expanding territorial control through violence and the deployment of local criminal cells. Academic research has showed the connection amongst *"the progressive geographic diversification of violence in Mexico and the diversification of activities"* by criminal groups (Ibid, 9). A US Congress report on the Mexican security situation stated that crime groups *"have splintered and diversified* their criminal portfolios adding *extortion, kidnapping, auto theft, oil smuggling, human smuggling, retail drug sales, and other illicit enterprises"* (Beittel 2018, 1).

During my fieldwork interviews one argument was recurring: the kingpin and anti-drug strategies implemented during Calderón and EPN had been crucial in pushing criminal groups into fuel trafficking. State enforcement has played an important role in the behaviour of criminal networks by increasing or decreasing the costs of engaging in certain illegal activities. Enhanced possibilities of repression increase those costs. Enforcement has led criminal groups to relocate, pushing diversification of illegal enterprises as they migrate; a phenomenon known as the *"cockroach effect"* (Durán-Martínez 2018, 22).

Increased enforcement has led to the formation of more groups, each diversifying their sources of income, producing *"a highly complex multipolar criminal market"* (Felbab-Brown 2019, 10). In this context, market-sharing agreements and deterrence are almost impossible. The state's role is relevant, but not absolute. Market incentives of criminal enterprises matter as well: potential profits can shape the way criminal actors compete, cooperate and diversify into new illicit activities. The sole causality behind the effects of the kingpin strategy is contentious beyond the incentives offered by black markets, criminal groups' internal dynamics also play a role.

Criminal Dynamics: Fragmentations and Alliance Formation

In Chapter 2 the factors pushing criminal fragmentation were developed, giving criminal dynamics their importance without neglecting the influence of the state. It was established that fragmentations can occur because of (1) the breakdown of alliances, (2) a group gaining independence and (3) internal succession processes. Alliance-making is another form of reconfiguration linked to fragmentation that can occur when (1) weakened groups ally, (2) to confront a common enemy or (3) to pursue territorial control or provisional common interests (Atuesta and Pérez-Dávila 2018, 240). All these circumstances can operate in conjunction. In this section we will explore some examples for these scenarios to further understand the fragmentation beyond the state's kingpin strategy.

La Familia Michoacana is an example of an alliance between local groups to confront a common enemy. This took place in 2006, when the Zetas expanded into Michoacán to take control of the Port of Lázaro

Cárdenas, a strategic criminal asset that enabled access to chemical precursors for methamphetamine production. A case of fragmentation triggered by heterogeneous factions is exemplified by the Zetas and the GC. After the leader of the GC, Osiel Cárdenas was detained, the Zetas broke away from the organization. The Zetas were conceived as an armed wing of the GC but since their inception operated with their own military rules and independent structure (Atuesta and Pérez Dávila 2018). This autonomy facilitated their separation from the GC, starting a turf war between these groups. After the capture of Cárdenas, subsequent leaders were either killed or detained. The sustained fall of GC leaders also influenced the emancipation of the Zetas in 2010. Another factor playing a role in the Zetas–Gulf fragmentation was the territorial expansion that the Zetas pursued independently from the GC, which led to conflicts in the distribution of profits. An additional reason was the death of a close operator of the Zetas' leader Heriberto Lazcano at the hands of the GC (Gómez 2010). These examples demonstrate how multiple factors can combine leading to criminal fragmentation.

These are a few examples of fragmentation and cooperation dynamics reconfiguring the Mexican criminal landscape towards an increased diversification and pluralism. Having explored cases of the dynamics driving fragmentation of criminal groups, we will now explore how fragmentation has led towards illegal market diversification in general and into the MFBM in particular.

Diversification Towards New Criminal Markets

As criminal groups fragment, they diversify their income sources. Criminal groups formed from fragmentations tend to engage in illicit activities outside transnational drug trafficking to access resources and increase their resilience against state repression and violent illicit competition. Larger networks also pursue capturing revenues of new profitable black markets. Groups of all sizes are involved in this diversification trend enhanced by the fragmentation cycle that began in Mexico in the mid-2000s. Currently, the Mexican illicit landscape is dominated by organizations that depend on engaging in different criminal activities through territorial control. This in turn has created a vicious trend of constant territorial disputes accompanied by growing violence pushing the need for further resource extraction, more territorial expansion and increasing aggression. These dynamics are further enhanced by criminal fragmentation and influenced by other factors like economic growth. Economic growth generates new markets for diversifying criminal networks including commercializing illegal goods and services (narcotics, prostitution) and predatory practices like extortion, kidnapping and extractive activities (Garzón 2014). Urban centres concentrate economic and human resources appealing to criminal networks that can enable diversification (informal economies, extortion) and exploit money-laundering opportunities inaccessible in rural areas (Durán-Martínez 2018, 6–7). Urban areas have become strategic for Mexican criminal groups because they guarantee access to goods and services relevant for their operations (vehicles, financial and communication services) (Fuerte Celis et al. 2018, 9). Fuel trafficking in Mexico is thriving in urban areas with vibrant economic activity. The municipalities displaying the highest levels of pipeline extraction between 2009–2016 were in areas with urban populations and four of them belonged to a larger metropolitan area (Table 4.1).

Fuel trafficking requires high-way infrastructure to transport fuel out of energy installations. The presence of highways and connectivity were recurring themes during fieldwork, and they are key to understanding the location of fuel trafficking operations. Highways and transport infrastructure are near urban centres, offering resources for *"the illegal removal and commercialization of stolen fuel"* (EnergeA and Grupo Atalaya 2017, 49).

This research developed how fragmentation and competition created several motivations for diversification amongst Mexican criminal actors. What is yet to be explored is the origins of the diversification trend and how it led to fuel trafficking. And to explain this we must turn our attention to the Zetas. In the following sections we will explore the model of black-market diversification this group pioneered and their role in branching out into fuel trafficking beyond the Burgos Basin.

The Zetas: Pioneering Black-market Diversification

The expansion of sophisticated criminal networks into fuel trafficking in Mexico is linked to the Zetas, a group whose irruption represented a turning point for Mexico´s security crisis. By 2013 security analysts warned that this group *"managed to diversify its sources of revenue. Rather than concentrating on trafficking drugs, the Zetas' portfolio includes everything from piracy, extortion, kidnapping, and migrant smuggling to theft*

Table 4.1 Municipalities with more illegal pipeline extraction points 2009–2016

Municipality	Population type	Metropolitan municipality	Metropolitan area
Acajete, Puebla	Urban	Yes	Puebla—Tlaxcala
Acatzingo, Puebla		No	X
Altamira, Tamaulipas		Yes	Tampico
Cárdenas, Tabasco		No	X
Culiacán, Sinaloa		No	X
González, Tamaulipas		No	X
Huimanguillo, Tabasco		No	X
Irapuato, Guanajuato		No	X
Mocorito, Sinaloa		No	X
Palmar de Bravo, Puebla		No	X
Quecholac, Puebla		No	X
Salamanca, Guanajuato		No	X
Salvador Alvarado, Sinaloa		No	X
Silao, Guanajuato		Yes	León
Tala, Jalisco		No	X
Tepeaca, Puebla		No	X
Tlajomulco de Zúñiga, Jalisco		Yes	Guadalajara

Source EnergeA and Grupo Atalaya (2017, 49)

from oil pipelines and levying taxes on other criminal organizations" (Correa-Cabrera 2017, 60). These activities generated higher revenues for the Zetas than narcotics.

By pioneering enterprise diversification, the Zetas radically changed the model of what territorial control represented for criminal networks, transcending the historic significance of narcotics. This was achieved by adopting a criminal franchise model with a loose network structure exploiting different illegal activities (Bailey 2011). This allowed the extraction of profits from subnational enclaves, while increasing organizational resilience against growing state enforcement. With this system the group engaged in large-scale drug smuggling, operating in Colombia, Venezuela, the US, Africa and Europe (Correa-Cabrera 2017). They developed a widespread extortion system, engaged in local drug dealing, human trafficking for *"labour, sexual exploitation and body organ harvesting"* (Ibid, 81).

The Zetas were the first group to link the routes and activities of drug smuggling, human trafficking and kidnapping, a response to underused illicit resources. As Deville (2013) argues, diversification of criminal enterprises *"allow illicit networks not only to further utilize their existing drug trafficking supply chain but also to further increase their political influence as well as their profits"* (69). A further example are the stash houses where human traffickers held migrants hostage while their "transport fees" were paid to the Zetas.

This diversification was complemented with disruptive practices. The Zetas aggressively expanded their territory, breaking the boundaries of the *plaza* system. To achieve this, they conducted brutal killings (displaying beheadings publicly) and militarizing their armed cells. This model changed the paradigm of criminality in Mexico. The Zetas were eager to participate *"in any enterprise that would generate profit"* (Del Bosque and Ulloa 2013). Other criminal actors followed: networks like *Familia Michoacana*, Knights Templar and *Cartel Jalisco Nueva Generación* (CJNG) (Correa-Cabrera 2017, 77) adopted this practice of securing new sources of revenue through diversification.

The Zetas´ adoption of pioneering practices did not stop there. The organization combined aggressive diversification and territorial expansion with the co-option of state actors, a method the Zetas would become famous for (Fuerte Celis et al. 2018, 10). But this group would not stop their diversification at kidnapping, extortion and human smuggling. They took their experience in the Burgos Basin and expanded into energy.

The Zetas as Fuel Traffickers

This segment will analyse the Zetas´ diversification into fuel trafficking and will explain how this group turned this illegal activity into a national-scale criminal phenomenon beyond local contexts. After 2010, reports pointed out to the Zetas´ innovative involvement in fuel trafficking. The location of the majority of detected illegal extraction points indicated the early involvement of this criminal group (Corcoran 2012). Pipeline extraction first appeared and consolidated in Zetas´ strongholds: Veracruz, Nuevo León, Puebla and Tamaulipas (Reinhart 2014, 753). Between 2007–2011 the Attorney General initiated 2,193 fuel-theft criminal investigations. From this total, the top state was Sinaloa (392 cases). What is interesting are the states that followed: Veracruz (308), Nuevo León (255), Puebla (210) and Tamaulipas (208) (Vega 2012). The Zetas strongholds concentrated 981 investigations, 150% more than Sinaloa. Most of those detained under these investigations identified as Zetas. The Zetas were pioneering the growth and geographic dispersion of fuel trafficking.

In 2013, PEMEX found 491 illegal extraction points in Tamaulipas, making it the state leading in this criminal activity with a yearly increase of over 180% ("Crece 200% el saqueo..." 2014). By 2014 analysts were warning about the Zetas attempts to infiltrate PEMEX. At the time, the Zetas and the GC consolidated a network of fuel distribution rivalling PEMEX in the north of the country. It was estimated that both groups controlled 15% of the gasoline in Tamaulipas and that this criminal enterprise generated more revenues than narcotics and human trafficking (Gurney 2014).

What pushed the Zetas to diversify into this illegal market? Fragmentation was an important catalyst. Research argues that the Zetas and the GC got involved in hydrocarbons theft in Tamaulipas as a response to increased government repression and to strengthen their organizations financially after conflict exploded between them. In a moment of diminished operational capacity, it was rational for both groups to exploit local black markets rather than depending on transnational drug trafficking (Correa-Cabrera 2017, 200). The falling-out between the GC and the Zetas was evident in 2010 when messages left in execution sites *"began to be signed either by the Gulf Cartel or by Los Zetas"* (when they were previously signed *"Golfo-Zetas"*) (Atuesta and Pérez-Dávila 2018, 251). As the Zetas faced leadership loss, infighting and government repression, the group diversified increasingly into activities requiring less logistical complexity than international drug smuggling.

The Zetas opened fuel trafficking as a new thriving criminal market: lucrative, accessible and low risk, making it an ideal option for criminals pursuing enterprise diversification (Correa-Cabrera 2017, 200). Consequently, other criminal networks got involved in exploiting PEMEX´s installations and pipelines. According to a former intelligence official, the MFBM began its growth period with the involvement of large criminal groups like the Zetas. Prior to 2006, this black-market largely remained a local illicit enterprise controlled by specialized groups (Montalvo 2017).

The influence of the Zetas was such that it established operational methods for fuel-theft cells. As Ana Lilia Pérez (2018) told me: fuel traffickers *"have copied the Zetas model. They use "stakes"* (jargon for criminal armed cells) *and "hawks"* (on the ground intelligence providers) *and install operational and surveillance fences; all these strategies were originally implemented by the Zetas"*. Stealing and commercializing stolen fuel was complemented with extorting PEMEX workers, suppliers and service providers. The Zetas instated a new criminal market with fuel trafficking, but this was one part of a larger transformation. With their diversification and their incursion into new criminal enterprises the Zetas altered Mexico's criminal landscape radically; setting the conditions for a new context in which the MFBM would consolidate. But before explaining this recently relevant criminal context, we must explain the conditions that make Mexico a thriving criminal hub.

Mexico As a Criminal Hub

The presence of macro-criminal networks and illicit enterprises of all sizes are not phenomena connected to a lack of development or the nonexistence of state presence. Still, it has become commonplace to connect criminal incidence to failed states and underdevelopment. This could not be further from the truth and, though exclusion undoubtfully plays a role in the criminality problem in Mexico, the 15th largest world economy, with developing modern industrial and service sectors, paints a more nuanced scenario between development and the presence of multiple criminal networks. This is because these groups rely on some *"level of infrastructure and services"* and that in reality, *"many of the dominant hubs in the illicit global economy can be found in the major cities of relatively coherent states"* (Radden Keefe 2013, 100).

Criminal networks are motivated by profits and require certain levels of governance and infrastructure to engage in high-margin operations. Therefore, criminal groups' ideal locations for stable illegal enterprises combine the benefits of a *"functioning state, such as modern infrastructure and communications, a banking system, and enough rule of law to make life generally predictable"* (Ibid, 102). Radden Keefe (2013) defines these locations as *"criminal hubs"*. Mexico fits this definition; it is prosperous enough country whose state functions enough to sustain an exports-based economy and an important destination for foreign investment.

Esparza (2014) argues that one of the reasons behind the Sinaloa Cartel being such a powerful criminal network is the presence *"high-capacity highways"* that connect Mexico's northern region to the state of California. Additionally, the ports of Mazatlán, Manzanillo, Lázaro Cárdenas and Salina Cruz give access to the trade routes of the Pacific, while both coasts are connected by the Mazatlán-Matamoros superhighway. This represents a *"unique logistics system in the world"* (Ibid) that gives Mexican criminal networks a competitive advantage. These elements, combine with the state´s lacking security and law enforcing capacities and the country´s unique geographic location, make it a thriving criminal hub.

Other elements play a part in understanding Mexico as a criminal hub. Criminal networks find in Mexico´s informal economy, concentrating over 55% of the employed population (Levy 2018, 94), a critical component to sustain criminal operations because of its reliance on cash and the flow commodities in an unregulated space, allowing profits to be laundered with ease (Radden Keefe 2013, 106). The semi-functioning Mexican state also offers opportunities of widespread co-option to protect illicit enterprises. It has been observed, in Mexico and Latin America, that profitable subnational criminal economies offer the opportunity of sustenance for multiple criminal groups. Within these territories, ports, unmonitored areas, highways, energy installations and urban areas with informal economies are the most sought-after illicit assets (Yashar 2018, 125). These elements making Mexico a criminal hub have spawned one of the most competed criminal landscapes in the world.

Mexico´s Criminal Landscape

It is important to develop the context in which Mexican fuel trafficking is thriving. Since 2008–2010 a criminal fragmentation trend consolidated in Mexico, setting the criminal underworld in a path of increasing complexity. Disintegration of illicit organizations is dominant amongst Mexico´s criminality and has marked the end of large drug cartels to give way to smaller and diversified criminal groups (Asmann 2019). Currently, there are multiple smaller criminal networks in Mexico who coexist along stronger groups with broader territorial presence, for example *Cártel Jalisco Nueva Generación* (CJNG) (Pérez Caballero 2018).

The CJNG was born from an alliance between groups of *La Familia Michoacana*, that became independent in 2014 after the death of its leader, and an offshoot of the Sinaloa Cartel known as *Los Mata Zetas* (Atuesta and Pérez-Dávila 2018) that fragmented from Sinaloa in 2013 (Jones 2018, 21). Insight Crime estimates this group has presence in 20 states (Beittel 2018), while Mexican security institutions assess that presence in 25 states (Muedano 2018). It is the first criminal network to simultaneously have dominance in key states in both of Mexico's coasts with active participation in fuel trafficking and other emerging markets (Guerrero Eduardo 2018a).

The rise of the CJNG reflects a scenario in which macro-networks coexist with smaller organizations. Large networks in the Tierra Caliente region (Mexico's southwest, a foundational stronghold of the CJNG) have the capacity to engage in translational drug trafficking, based on the opiate and methamphetamine trade enabled by their control over opium-producing areas and the ports of entry for synthetic precursors from the Pacific (Stewart 2018). These larger networks, which also exploit local criminal markets, share the Mexican territory with smaller networks that have no capacity to engage in transnational drug trafficking and depend on more local-predatory enterprises like extortion, kidnapping, cargo-theft, carjacking and fuel-theft to obtain revenues to sustain their operations (Ibid). This diverse and fragmented criminal environment is linked to increasing levels of violence in the country. Homicides reached a record-high in 2017, while in 2018 homicides increased in 27 states (a total of 32) and in 15 of them projections pointed that at the end of that year historical records would be broken (Ángel 2018).

This situation is constantly changing, as criminal competition and state enforcement make the illicit underworld unstable. This makes determining accurately the involved actors dauntingly difficult. US analysts estimate that in 2006 there were six relevant criminal groups in Mexico. By 2012 that number grew to over ten and by the end of 2014 there were more than twenty criminal groups (Beittel 2018, 26). Mexican authorities have stated that six major criminal groups and eighty smaller cells were operating across Mexico in 2018 (Muedano 2018). To accommodate this mercurial diversity, Stratfor Intelligence has advocated for categorising criminal actors in Mexico by areas of influence[1] (Reed 2015). What these efforts show is how complex Mexico's criminal underworld has become. A similar attempt must be made to elucidate the criminal networks are participating in the MFBM. This research proposes that large criminal

---

[1] These areas of influence are: "Sinaloa Region" occupying Mexico's northwest states that have access to the Pacific, the "Tierra Caliente" in the southwest Pacific coast stretching inland into states from the Bajío and Centre areas of Mexico and the "Tamaulipas Region" occupying the states of the Mexican Gulf Coast from the northern border all the way into the Yucatán Peninsula in the south.

networks, specialized fuel trafficking networks and fuel criminal gangs all participate in the MFBM.

For the MFBM, control of extraction sources represents a strategic asset linked to criminal confrontation. These sources include distribution and marine terminals, refineries and pipeline sections. Control over these assets has been related to criminal territorial expansion and violent conflicts (Guerrero Eduardo 2018b). A second set of relevant assets are related to storage and distribution (warehouses, fuel containers, tanker trucks and privately-owned properties). Regarding transport these assets include marine vessels, trucks, tanker trucks and vans depending on the group's sophistication level. Finally, commercialization outlets range from highway sections, small businesses, gas stations and distribution to wholesale consumers.

Groups can confront one another in disputes for the MFBM in different enclaves. In some cases, alliances occur with smaller cells operating as sub-networks associated with larger groups. The following is an effort to clarify specialized fuel trafficking sub-networks affiliated with larger criminal networks. This is the result of crossing information between official, journalistic sources and reviewed with fieldwork informants and academic work, with the objective of making sense of the MFBM and its operations.

Specialized Fuel Trafficking Networks Affiliated to Larger Criminal Groups

Currently, large criminal networks like the CJNG, the Sinaloa Cartel and the Zetas exploit the MFBM. According to Mexican security sources, the CJNG, the Sinaloa Cartel and (based on my research) the Zetas, along with nine specialized fuel trafficking networks control 60% of fuel-theft in four hotspot states: Hidalgo, Puebla, Guanajuato and Estado de México. In these states the MFBM generates yearly profits of $47 billion MXN ($2.4 billion USD) stealing 54,500 barrels of fuels daily (Dávila 2019). These specialized networks demonstrate the levels of diversification Mexican criminal networks have exhibited in relation to fuel trafficking (Table 4.2).

This method of establishing (or associating with) fuel trafficking groups was also mentioned during fieldwork. According to a journalist who investigated illicit activities in Guanajuato's Salamanca refinery, the

Table 4.2 Fuel trafficking specialized subnetworks

Fuel trafficking sub-network	Barrels of fuel stolen a day in pipelines	Area of operations	Criminal network affiliation
Los Téllez	6,900	Puebla / Estado de México / Tlaxcala	CJNG
El Bukanas	6,900	Puebla / Veracruz	Los Zetas
El Rapid-Inn	–	Puebla / Estado de México	Local Group
Los Talachas	7,800	Hidalgo	?
La Parka	5,800	Hidalgo	CJNG
Los Bárcenas	5,800	Hidalgo	Los Zetas / CJNG
Los Pelones	2,500	Guanajuato	Sinaloa
Santa Rosa de Lima	6,100	Guanajuato	Independent
Melchor Ocampo	3,800	State of Mexico	CJNG

Dávila (2019); Velázquez (2018); La silla Rota (2019); Dennis (2019)

CJNG operated with a model in which command structures were decentralized, implementing a franchise system that operated through cells. My respondent observed how the CJNG sent Jaliscians to Salamanca to develop these groups enlisting local criminals, while witnessing how fuel trafficking became a source of fast revenues for the group (Stargardter 2018).

Other sources reiterate this tendency towards forming fuel trafficking subnetworks. The governor of Hidalgo declared that cartels *"opened, like corporations, divisions for hydrocarbons"* ("Huachicoleros pugnan por..." 2019). This concurs with Correa-Cabrera's argument regarding the subsidiary-based diversification model pioneered by the Zetas. Under this model, cells operate like subsidiaries exploiting different criminal markets like drug trafficking, kidnapping, extortion, human trafficking, firearms smuggling, illegal logging, mining and hydrocarbons trafficking (2017, 56).

A similar argument has been made by Garay-Salamanca and Salcedo-Albarán (2016), affirming criminal groups can establish sub-networks. Specialization is a key characteristic in creating these groups: for example, sub-networks specialized in money-laundering, public official co-optation and so on. While researching the Zetas they concluded this group established *"criminal structures specifically responsible for the theft and trafficking of hydrocarbons, thus complementing other activities that report economic profit to the network"* (Ibid 119). In relation to the Burgos Basin trafficking operations of the mid-2000s, where both authors accessed court case documents, it was observed that out of a Gulf-Zetas macro-criminal network of 313 nodes, 53% of them were dedicated to narcotics while 8% to fuel trafficking (Ibid, 122). The size of this sub-network was small, unlike its economic impact: between 2007–2009 its operations led to approximate losses of $80 million USD ($1.5 billion MXN) (Ibid, 110). This shows how a small fuel trafficking sub-network can generate considerable revenues for a macro-criminal group.

Criminal diversification has been a driving force behind the expansion of the MFBM. Yet, its scale cannot be understood without the co-option of PEMEX and other state and non-state grey actors.

Co-option of Grey Actors

Through the widespread participation of technicians, engineers, union representatives and officials, criminal networks accessed PEMEX's installations, acquired expertise, retrieved transport information, hid the scale of their operations and accessed official resources. The co-option of PEMEX led to a Co-opted Institutional Reconfiguration (CItR) in which the SOE reproduced and legitimised illegal procedures benefiting illicit actors. The array of implicated agents and the scope of their operations (including billions in stolen resources, massive money-laundering, sophisticated façade companies, diverse illegality camouflaging methods and far-reaching manipulation of institutional procedures) makes this a case of CItR sustained through process of Systemic Macro-corruption. This is a central phenomenon behind the expansion of the MFBM. In this section, the case of the CItR of PEMEX through the co-option of grey actors will be developed. Individuals involved in distribution and money-laundering will also be analysed.

PEMEX: A Breeding Ground for Co-option

Illicit practices have been tolerated in PEMEX since the company's inception. This made this SOE an ideal entity for criminal co-option. Fraudulent practices permeated from all instances, from the low-level workforce to the union leaders and public officials with directive positions. A recurrent incident in the Tula refinery exemplifies this. Employees of the refinery routinely smuggled small quantities of hidden fuel out of the refinery. Workers described this practice as an unwritten agreement since their superiors stole bulk amounts of fuel (Pérez 2011). Ana Lilia Pérez (2018) explained that workers frequently sold PEMEX equipment (uniforms), while tanker truck drivers regularly sold fuel from their cargo for a profit. If workers engaged in unlawful acts and small-scale theft, the directives of the company also have a long history of corruption scandals. These are too many to document, but some examples include:

Pemex officials were accused of diverting over $1 billion MXN ($51 million USD) of public funds to the presidential campaign of the PRI candidate Francisco Labastida in 2001. Between 2003–2012 more than 100 contracts worth $11.7 billion USD presented irregularities, including ghost employees, payments for non-existent services and conflicts of interest (Toledo 2016, 3). In 2011, PEMEX paid $9 million USD to transport an oil rig from the Persian Gulf to the Gulf of Mexico. It was later revealed that the rig was never delivered (Ibid). In 2016 the Brazilian company Odebrecht paid $10.5 million USD in bribes to Mexican officials. The main broker involved in this scandal was Emilio Lozoya, the Director of PEMEX and a member of President Peña Nieto´s inner circle (Ahmed 2018). In all these cases impunity imposed itself, showing that corruption in PEMEX is an omnipresent scourge affecting the company at all levels. In this context the criminal operators of the MFBM would find a breeding ground to co-opt multiple grey actors within PEMEX and trigger its expansion.

Oil Workers: Crucial Grey Actors

The growth of the MFBM since 2011 required the involvement of state-actors, particularly PEMEX´s unionized workers and non-unionized employees. Obrador´s presidency described Mexican fuel trafficking as a large-scale fuel-theft and distribution scheme (López Obrador 2018). Fuel trafficking within PEMEX has also been defined as a parallel industry involving criminals, unionized workers, employees, contractors, businessmen, pipeline tappers, transporters, ship captains and crews (Pérez 2018). All these actors coordinate in stealing and selling hydrocarbons at a large-scale: approximately 9 million litres of fuel stolen daily, the equivalent of 600 carrier tanker trucks with a 15,000-litre capacity each representing profits of $200 million MXN ($10,297,800 USD) (López Obrador 2018).

In 2017, four individuals were detained near the Salamanca refinery in Guanajuato who were carrying *"weapons, army uniforms, technology to detect pipelines and drilling devices"* (Televisa 2018). One of the detainees was a PEMEX worker and another was a former employee. The worker, who was unionized, was freed and PEMEX couldn't fire him. This is one case of many. Moreover, it was revealed PEMEX had a blacklist of 130 workers detained for participating in fuel trafficking, including drivers, pumping operators, watchmen and engineers. Most were detained in Tamaulipas, Guanajuato, Puebla, Sinaloa and Nuevo León. 70% of the blacklisted employees belonged to *PEMEX Refinación* (renamed *PEMEX Logística*), the subsidiary in charge of fuel storage and transport (Solera 2017). Widespread corruption practices developed within PEMEX, and the union leadership is no exception.

The Union and the Plunder

PEMEX workers of all levels participate in a complex structure dedicated to fuel-theft (Pérez 2018a). The SOE's workers union (STPRM), has long been denounced for its collusion. In 2017 journalist Loret de Mola published in a column, after interviewing PEMEX insiders, that within the SOE it was an open secret that the union leaders were involved in fuel-theft. In 2015 the leader of the dissident worker's movement, Jorge Fuentes García, denounced the leadership of the STPRM for profiting from fuel trafficking. Besides denouncing the union leader Carlos Romero Deschamps, Fuentes García mentioned Luis Ricardo Aldana Prieto, head of the Union Sect. 40 based in Mexico City. According to Fuentes, Aldana Prieto was involved in fuel trafficking in Puebla, Querétaro and Hidalgo, all fuel trafficking hotspots. Besides Aldana, other regional union leaders were singled out for being complicit in fuel trafficking in Tamaulipas, Guanajuato, Veracruz, Tabasco, Estado de México and Hidalgo (Medina 2015). The STPRM has also been accused of having links with the Zetas (Correa-Cabrera Guadalupe 2017, 83). Another dissident labour group filed a criminal complaint in 2019 accusing Union Sect. 30 leader, based in Veracruz, for providing information to the Zetas to exploit pipelines (Vergara 2019).

White Collars and Bureaucrats

At the white-collar level, balances of PEMEX's wholesale operations were manipulated, and false transactions were validated. This happens at the beginning of the refining process: each refinery receives an amount of crude oil and raw materials according to its processing capacity. These amounts are measured and registered when they arrive to these installations through pipelines or tank trucks, and production levels must match them. When receiving these supplies, the established amounts are altered, allowing for unreported fuel to be stolen. As mentioned earlier, PEMEX's Director admitted in 2019 that the oil company does not have effective control mechanisms to ensure there are no differences between production and balances in refineries and distribution terminals (Romero 2019b), making these installations vulnerable to the MFBM.

Former PEMEX Director Emilio Lozoya concentrated in the Logistics subsidiary the transport of hydrocarbons. PEMEX Logistics became key for the SOE, having under its responsibility the operation of the pipeline network, 9 compression stations, 56 pumping stations, 16 tankers, 75 small vessels, 3,577 tank trucks and 511 tank cars. This subsidiary was being investigated by the Obrador administration, and three of its officials have been forcibly removed for their involvement in fuel trafficking.

Five workers of PEMEX's pipeline monitoring unit were removed from their posts and were investigated for ignoring detected extraction procedures, uncovering cases in which drops in pipeline pressure were ignored for 8–10 hours (*Animal Político* 2019). A high-level official that stands out is Army General Eduardo León Trauwitz, who was designated to lead the Strategic Security unit of PEMEX during EPN's administration. Trauwitz was responsible for overseeing the security of PEMEX's pipelines and other installations. In May 2019, the Attorney General accused Trauwitz of fuel-theft ("General León Trauwitz comparecerá" 2019).

Three former employees of the Strategic Security unit accused high-level officials who operated under the orders of Trauwitz, denouncing that these officials instructed personnel not to report illegal extraction points in pipelines in Tamaulipas, Nuevo León, Puebla and Veracruz. It was also revealed that a high-level official ordered PEMEX personnel to install an extraction point in Puebla in 2017. These accusations included PEMEX regional managers, some of the highest-level officials in the SOE's hierarchy (Barajas 2019).

The Distribution Network

Other important actors for the MFBM are individuals who collaborate distributing derivatives. They include businessmen, lawyers, financiers, elected representatives and gas station owners, constituting a group of state and non-state grey actors. This section will explore how the MFBM's distribution operates, developing the role of the grey actors involved. Distribution of stolen fuels can involve establishing façade companies with illegal resources to function as suppliers. An example of this involves the Gulf Cartel, which used a legally constituted company (*PetroBajío*) to channel hydrocarbons stolen in Altamira, Tamaulipas to buyers in Lázaro Cárdenas, Michoacán. This case illustrates the importance of façade companies for large-scale stolen fuel distribution (EnergeA and Grupo Atalaya 2017, 71).

"Gasoline wholesale requires gas stations, which are the only places where you can store it in large quantities. With diesel it is possible that industrialists buy it in bulk, gasoline needs stations" (Dante 2018). This was the response made by a former *PEMEX Refinación* official to a question I asked regarding the distribution of the MFBM. Facts back his affirmation: a small gas station has a storage capacity of 15,000–17,000 litres, while a large station can store 40,000 litres (Barragán 2019). Journalistic sources cited PEMEX insiders corroborating the involvement of gas stations in the trade of stolen fuels (Jiménez 2012). A statement in 2015 by the Mexican Association of Gasoline Entrepreneurs said 20% of the market was under the control of criminal groups (Correa-Cabrera 2017, 217).

In January 2019 tax authorities revealed, after investigating the purchases and sales of the 12,000 gas stations in Mexico, that 194 stations had severe taxing inconsistencies linked to fuel trafficking totalling $3.2 billion MXN ($167 million USD). 54% of this total ($88.7 million USD) was concentrated by gas stations in Mexico City, Estado de México, Michoacán, Tamaulipas, Baja California, Jalisco and Puebla (López Obrador 2019). Mexico's Financial Intelligence Unit presented five criminal proceedings for fuel trafficking involving fifteen gas station business groups. Another thirteen had pending proceedings.

One of these groups had dubious operations of $86.9 billion MXN ($4.5 billion USD), an amount equalling the annual budget of the government of Chiapas (Velasco 2018), and profits of $6.7 billion MXN ($350 million USD). These revenues were larger than the sum of the budgets of the National Human Rights Commission, the National Transparency Institute and the Telecommunications regulator for 2019 (Rodríguez 2019). This corroborates the importance of gas stations in fuel trafficking´s distribution network. But these installations can also play key roles as money launderers. Gas stations are used as façade companies to convert, conceal, or use illegal revenues and allow *"rapid cash turnover to blend funds from licit and illicit sources"* (Organization of American States 2013, 30).

Another source of demand of the MFBM are industrialists, who consume diesel at wholesale level at $14–$15 MXN/litre, while the legal price was $20 MXN/litre. Buyers are aware of the product´s illicit origin due to its cost (Dante 2018). Representatives of Mexico´s transport sector estimated that one third of the trucks dedicated to freight transport consume stolen diesel (180,000 vehicles out of 550,000) (Cruz 2017). Freight transport with the US is the main reason for Mexico´s northern region concentrating most diesel-theft in the country (EnergeA and Grupo Atalaya 2017, 8). It is true that the MFBM consolidated a large-scale distribution scheme, in which most of the stolen hydrocarbons are sold within Mexican territory through a sophisticated supply network enabled by state and non-state grey actors.

Grey Actors and the Co-opted Institutional Reconfiguration of PEMEX

The importance of grey actors in the MFBM is central. Unionized workers, leaders, employees, officials and businessmen provide criminal networks strategic institutional and technical resources enabling the expansion of this enterprise. A recurring resource provided by grey actors is money-laundering. This activity *"involves introducing illicit funds into the financial system, distancing the funds from the criminal(s), and converting the money into legitimate business earnings"* (Malm and Bichler 2013, 366) and is a strategic resource for the sustained functioning of criminal networks (Garay-Salamanca and Salcedo-Albarán 2016, 33). It connects legal economic agents to unlawfulness and generates grey zones in which legitimate actors can partake in criminal operations (Organization of American States 2013, 6). Money-laundering enables criminality to infiltrate *"different spheres of society"* (Ibid, 29) and co-opt state and non-state actors.

Co-opting PEMEX is critical. The routine actions behind fuel-theft in the SOE´s refineries, terminals and pipelines require a robust presence of low and mid-level PEMEX grey actors who provide criminal networks with access, expertise, information, transport, storage, documentation and enable illegality camouflaging. Without them, the daily operations that together form the MFBM could not occur. Yet, the higher we go in the command chain the importance of grey actors increase. The involvement of high-level officials and union leaders is crucial for fuel trafficking. These actors, from their positions of power, can guarantee impunity by concealing the stolen resources. By sustaining impunity, they allowed the MFBM to expand massively making them *"network(s) extenders"* (Williams 2001, 81). The participation of powerful private and public agents is an indispensable condition for criminal enterprises to reach macro-criminality status, which is the capacity to operate in illicit markets, in the legal economy and in the contact points in-between (grey zones) (Garay-Salamanca and Salcedo-Albarán 2016, 24 and 153).

The expansion of fuel trafficking reveals the CItR of PEMEX, through a process of Systemic Macro-corruption. The number of state and non-state actors involved, and the sophistication of their operating methods make this a case of Systemic Macro-corruption. The omnipresence of grey actors in PEMEX´s hierarchy generated sustainable agreements for profits *"beyond a single bribe"* (Garay-Salamanca et al. 2018, 31). Simultaneously, an instrumental capture of PEMEX diverted its institutional purposes for egotistic benefits. Through PEMEX´s Systemic Macro-corruption, fuel-theft evolved into a consolidated black-market that inserted itself as a parallel structure within the PEMEX Production Chain, interfering in Industrial Transformation, Logistics and Commercialization (Illustration 4.1).

The consolidation of the MFBM did not only occur due to grey actors. The market also became increasingly profitable for criminal networks to exploit, partly due to soaring fuel prices.

FUEL PRICES INCREASES IN MEXICO

The growth of the MFBM was parallel to significant increases in fuels prices (Fig. 4.1).

Between 2011–2018 *Magna* gasoline increased 135.4%, *Premium* Gasoline increased 147% in the northern border (144.4% in the rest of the

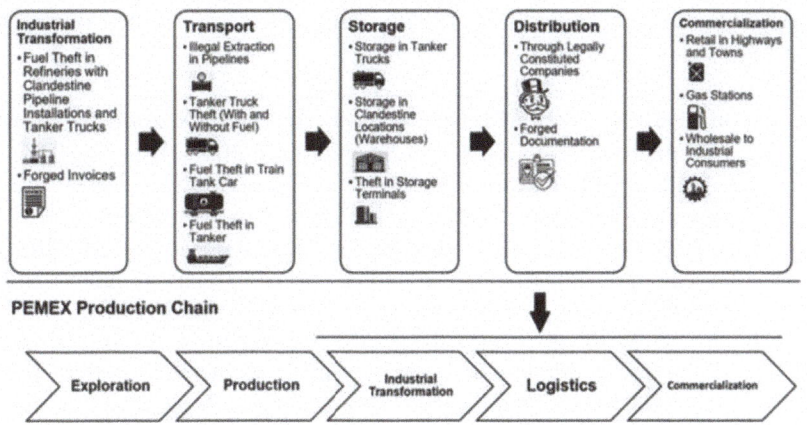

Illustration 4.1 The Mexican Fuels Black Market (MFBM) (Elaborated by the author)

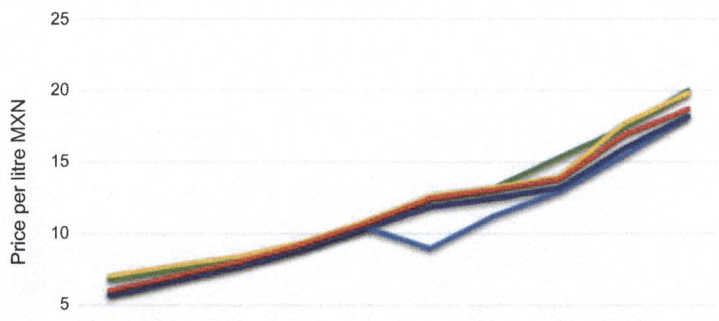

	2009	2010	2011	2012	2013	2014	2015	2016	2017	2018
Magna (NB)	5.75	6.73	7.73	8.94	10.41	8.98	11.26	12.95	15.43	18.2
Premium (NB)	6.77	7.46	8.1	9.08	10.56	12.6	13.19	15.32	17.34	20
Magna (RC)	5.73	6.73	7.73	8.94	10.41	11.88	12.45	13.18	15.93	18.2
Premium (RC)	7.06	7.76	8.41	9.4	10.89	12.6	13.19	13.97	17.66	19.8
Diesel	6.02	7	8.01	9.24	10.72	12.45	13.02	13.8	16.85	18.7

Fig. 4.1 Evolution of Hydrocarbon Prices in Mexico 2009–2017 (per litre) *NB (Northern Border) RC (Rest of the Country) (Secretaría* de Energía 2018, *44)*

country) and diesel prices grew 133.4%. These increases can be explained through various factors:

I. Since 2004 Mexican oil production declined 4% annually from 3.38 million barrels a day to 1.95 million in 2017 (Estrada 2018). Falling production mixed with declining refining capacity and growing internal demand increased dependency on imports.
II. In 2017 68% of gasoline consumed in Mexico was imported. These imports, combined with the depreciation of the Peso against the US Dollar, translated into costs of $24 billion USD (Ibid).
III. The prices of oil in global markets declined in 2014 due to an oversupply generated by fracking and Saudi Arabia flooding the global oil market. During the Fox–Calderón administrations the higher prices of the oil barrel were used to subsidize fuels. This changed under EPN who increased taxes (IEPS) to compensate for these losses (Fig. 4.2).

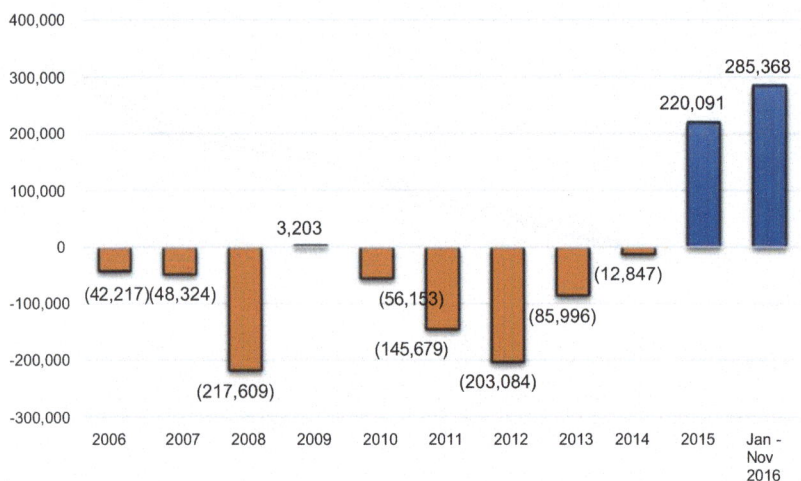

Fig. 4.2 From Subsidies to Taxation: Evolution of the IEPS during the Calderón and the Peña Nieto Administrations *(Elaborated by author with information of Moreno 2017)*

IV. In 2016 47% and 38% of the prices of Magna gasoline and diesel respectively corresponded to taxes (Pérez García 2018, 14), increasing prices due to the government's dependence on fuel revenues.
V. In January 2017, the EPN administration announced a 20% increase in the prices of fuels as the fiscally unsustainable gasoline subsidy was removed, and prices were adjusted to international standards.

Price increases represented a key incentive for criminal networks to expand their involvement in the MFBM. Between 2011–2018 *Magna* gasoline prices increased 97% (adjusted for inflation). *Premium* in the north increased 108.4% (106% in the rest of the country). Diesel prices increased 95%. This made the MFBM extremely lucrative and created a growing demand for illegal fuels, the "*propelling force behind illegal market exchanges*" (Beckert and Dewey 2017, 3) (Fig. 4.3).

These increases made fuel trafficking more profitable than other enterprises like human or drug trafficking (Correa-Cabrera 2017, 82). A fuel

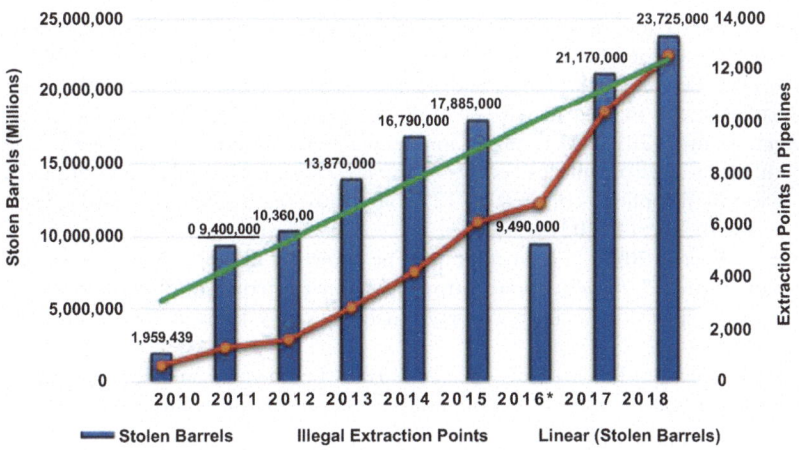

Fig. 4.3 Stolen Barrels vs Extraction Points (2010–2018) *(Sánchez 2012; EnergeA and Grupo Atalaya 2017, 344; Romero 2019a; PEMEX: "Reporte de tomas clandestinas")* *Official data for stolen barrels reported in 2016 did not include pipeline losses

trafficker corroborated this: "*The more they increase, the more they pay to us. It's better for us. The more they pay us, the more each worker earns ... and the more for [PEMEX workers involved in illegal extraction]*" (Osorno 2017).

Bibliography

"Cinco personas a cargo de monitorear ductos de Pemex fueron despedidas y denunciadas, dice la secretaria de Energía." *Animal Político*. January 9, 2019. https://www.animalpolitico.com/2019/01/monitoreo-pemex-despidos-rocio-nahle/ Date accessed: July 14, 2021.
"Crece 200% el saqueo de ductos en Tamaulipas." *Milenio Digital*. February 2, 2014. https://www.milenio.com/estados/crece-200-el-saqueo-de-ductos-en-tamaulipas Date accessed: July 14, 2021.
"General León Trauwitz comparecerá hoy para conocer acusaciones de FGR." *Milenio Noticias*. May 31, 2019. https://www.youtube.com/watch?v=vvckvc l3I5Q Date accessed: July 14, 2021.
"Huachicoleros pugnan por plazas en Hidalgo, dice Fayad." *Criterio Hidalgo*. February 13, 2019. https://www.criteriohidalgo.com/noticias/huachicoleros-pugnan-por-plazas-en-hidalgo-dice-fayad Date accessed: July 14, 2021.
Ahmed Azam. 2018. "Mexico Could Press Bribery Charges. It Just Hasn't." *The New York Times*. June 11. https://www.nytimes.com/2018/06/11/world/americas/mexico-odebrecht-investigation.html Date accessed: July 14, 2021.
Ana Lilia Pérez. 2018. (Independent journalist and author) interviewed by Samuel León in Mexico City, August 23.
Ángel Arturo. 2018. "En 2018 los homicidios crecieron en 27 estados y en 15 alcanzaron niveles récord." *Animal Político*. December 28. https://www.animalpolitico.com/2018/12/homicidios-estados-niveles-record/ Date accessed: July 14, 2021.
Asmann Parker. 2019. "Fragmentation: The Violent Tailspin of Mexico's Dominant Cartels." *Insight Crime*. https://www.insightcrime.org/news/analysis/violence-spikes-criminal-groups-fragmment-mexico/ Date accessed: July 14, 2021.
Atuesta Laura and Pérez-Dávila Yocelyn. 2018. "Fragmentation and cooperation: the evolution of organized crime in Mexico." *Trends in Organized Crime* 21: 235–261.
Ayling Julie. 2009. "Criminal Organizations and Resilience." *International Journal of Law, Crime and Justice* 37: 182–196.
Bailey John. 2011. "What do the Zetas and McDonald's Have in Common?." *Insight Crime*. December 5. https://www.insightcrime.org/news/analysis/what-do-the-zetas-and-mcdonalds-have-in-common/ Date accessed: July 14, 2021.

Barajas Abel. 2019. "Exhiben militares ordeña de General." *Reforma*. April 7.
Barragán Almudena. 2019. "¿Las gasolineras sin combustible están cerradas por vender huachicol?." *El País*. January 10. https://verne.elpais.com/verne/2019/01/10/mexico/1547094217_684823.html Date accessed: July 14, 2021.
Beckert Jens and Dewey Matías. 2017. "The Social Organization of Illegal Markets." In *The Architecture of Illegal Markets: Towards an Economic Sociology of Illegality in the Economy*, edited by Jens Beckert and Matías Dewey, 1–38. Oxford University Press.
Beittel June S. 2018. "Mexico: Organized Crime and Drug Trafficking Organizations." *Congressional Research Service*, 1–29. July 3.
Corcoran Patrick. 2012. "Oil Theft is Big Business for Mexican Gangs." *Insight Crime*. March 20. https://www.insightcrime.org/news/analysis/oil-theft-is-big-business-for-mexican-gangs/ Date accessed: July 14, 2021.
Correa-Cabrera Guadalupe. 2017. "Los Zetas Inc.: Criminal Corporations, Energy, and Civil War in Mexico," 1–341. University of Texas Press.
Cruz Serrano Noé. 2017 "Usan huachicol 180 mil vehículos de transporte." *El Universal*. July 5. http://www.eluniversal.com.mx/articulo/cartera/economia/2017/07/5/usan-huachicol-180-mil-vehiculos-de-transporte Date accessed: July 14, 2021.
Dante Yamil San Pedro (Energy Advisor for Mexico's Business Coordinating Council (CCE) and former PEMEX official) interviewed by Samuel León in Mexico City, August 10, 2018.
Dávila Patricia. 2019. "El Mapa del Huachicoleo." *Proceso*. January 27.
Del Bosque Melissa and Ulloa Jazmin. 2013. "Bloodlines: How the Scion of a Texas Horse Racing Empire Became an Informant on Mexico's Most Feared Cartel." *Texas Observer*. August 7. https://www.texasobserver.org/bloodl ines-how-the-scion-of-a-texas-horse-racing-empire-became-an-informant-on-mexicos-most-feared-cartel/ Date accessed: July 14, 2021.
Deville Duncan. 2013. "The Illicit Supply Chain." In *Convergence: Illicit Networks and National Security in the Age of Globalization*, edited by Michael Miklaucic and Jacqueline Brewer, 63–74. National Defense University Press.
Durán-Martínez, Angélica. 2018. *The Politics of Drug Violence: Criminals, Cops and Politicians in Colombia and Mexico*, 1–299. Oxford University Press.
Edmundo Velázquez (Journalist and Director of newspaper *Página Negra*) interviewed by Samuel León in Mexico City, August 20, 2018.
EnergeA and Grupo Atalaya. 2017. "Estudio para analizar la problemática de seguridad física en las instalaciones del sector hidrocarburos y emitir recomendaciones para el reconocimiento de costos por concepto de seguridad que la Comisión Reguladora de Energía lleva a cabo en sus procesos de revisión tarifaria," 1–368.

Estrada Liliana. 2018. "La causa detrás de la importación de gasolinas." *Animal Político*. February 15. https://www.animalpolitico.com/blogueros-inteligen cia-publica/2018/02/15/importacion-gasolinas/ Date accessed: July 14, 2021.

Felbab-Brown Vanda. 2019. *Mexico's Out-Of-Control Criminal Market*, 1–29. The Brookings Institution.

Fuerte Celis María del Pilar, Pérez Lujan Enrique and Cordova Ponce Rodrigo. 2018. "Organized crime, violence, and territorial dispute in Mexico (2007–2011)." *Trends in Organized Crime* 22: 188–209.

Gabriel Stargardter (Journalist for Reuters) interviewed by Samuel León in Mexico City, August 9, 2018.

Garay-Salamanca Luis Jorge y Salcedo-Albarán Eduardo. 2016. "Macro-Criminalidad: Complejidad y Resiliencia de las Redes Criminales." *iUniverse*, 1–191.

Garay-Salamanca, Luis Jorge, Eduardo Salcedo-Albarán, Macías Fernández Guillermo, Diana Santos Cubides, and Nathalia Guerra Villamizar. 2018. *Macro-Corruption and Institutional Co-optation: The "Lava Jato" Criminal Network*, 13–209. Bogota: Vortex Foundation.

García Dennis. 2019. "La PF, tras empresas ligadas al huachicol." *La Jornada*. January 24. https://www.jornada.com.mx/2019/01/24/politica/007n2pol Date accessed: July 14, 2021.

Garzón, Juan Carlos. 2014. "From Drug Cartels to Predatory Micro Networks: The 'New' Face of Organized Crime in Latin America." In *Reconceptualizing Security in the Western Hemisphere in the 21st Century*, edited by Bruce M. Bagley, Jonathan D. Rosen, and Hanna Kassab, 117–131.

Gómez Francisco. 2010. "Ejecución de El Concorde Detonó Guerra en Tamaulipas." *El Universal*. March 7. http://archivo.eluniversal.com.mx/nac ion/176125.html Date accessed: July 14, 2021.

Guerrero Eduardo. 2018a. "La segunda ola de violencia." *Nexos*. April 1, 2018. https://www.nexos.com.mx/?p=36947 Date accessed: July 14, 2021.

Guerrero Eduardo. 2018b. "Seguridad, ¿hasta cuándo?." *Nexos*. January 1, 2018. https://www.nexos.com.mx/?p=35375&fbclid=IwAR3aZTKJ87 ci2GkCifcXeK-xSrSwPOJW7j0zdxO-XmSo40BvSvY4jlpbP2M Date accessed: July 14, 2021.

Gurney Kyra. 2014. "Mexico Criminal Groups Running Sophisticated Distribution Networks for Stolen Oil." *Insight Crime*. June 18. https://www.insigh tcrime.org/news/brief/mexico-criminal-groups-running-sophisticated-distri bution-networks-for-stolen-oil/ Date accessed: July 14, 2021.

Jiménez Benito. 2012. "Ubican con varillas tomas clandestinas." *Reforma*. January 29.

Jones Nathan P. 2018. "The Strategic Implications of the Cártel de Jalisco Nueva Generación." *Journal of Strategic Security* 11 (1): 19–42.

Levy, Santiago. 2018. "Esfuerzos Mal Recompensados: La Elusiva Búsqueda de la" Prosperidad en México, 1–317. Washington, D.C.: Banco Interamericano de Desarrollo.
López Obrador Andrés Manuel. 2018. "Presidential Press Conference." *Speech, Palacio Nacional*, December 27.
López Obrador Andrés Manuel. 2019. "Presidential Press Conference." Presentation, *Palacio Nacional*, January 14, 2019.
Loret de Mola Carlos. 2017. "En Pemex también sospechan que el sindicato roba combustible." *El Universal*. February 6. http://www.eluniversal.com.mx/entrada-de-opinion/columna/carlos-loret-de-mola/nacion/2017/02/6/en-pemex-tambien-sospechan-que-el Date accessed: July 16, 2021.
Malm Aili and Bichler Gisela. 2013. "Using Friends for Money: The Positional Importance of Money-Launderers in Organized Crime." Trends in Organized Crime, 16 (June 16): 365–381.
Medina Francisco. 2015. "Líderes petroleros, involucrados en el robo a ductos: Jorge Fuentes García." Al Momento Noticias. February 11. http://diarioalmomento.com/lideres-petroleros-involucrados-en-el-robo-a-ductos-jorge-fuentes-garcia-e3zYye3DA.html Date accessed: July 14, 2021.
Montalvo Tania. 2017. "Así evolucionó el robo de combustible en México hasta provocar pérdidas millonarias." February 3, 2017. https://www.animalpolitico.com/2017/02/robo-combustible-mexico/ Date accessed: July 14, 2021.
Moreno José Miguel. 2017. "De Fox a Peña: Así se ha movido el precio de las gasolinas y por qué." *Imagen*. January 10. https://www.dineroenimagen.com/2017-01-10/82170 Date accessed: July 14, 2021.
Muedano Marcos. 2018. "Dominan 80 células del narco en México; operan seis cárteles." *Excélsior*. November 26, 2018. https://www.excelsior.com.mx/nacional/dominan-80-celulas-del-narco-en-mexico-operan-seis-carteles/1280724 Date accessed: July 14, 2021.
Muedano Marcos. 2019. "La Red de Municipios y Corrupción por Huachicoleo en Hidalgo." *La Silla Rota*. https://lasillarota.com/nacion/la-red-de-mun icipios-y-corrupcion-por-huachicoleo-en-hidalgo-huachicoleo-hidalgo-cuaute pec-tepeapulco/268333 Date accessed: July 14, 2021.
Organization of American States. 2013. "The Economics of Drug Trafficking." In *The Drug Problem in the Americas*. Organization of American States, 5–45.
Osorno Diego Enrique. 2019. ""Robar a Pemex es más redituable que ser zeta": Entrevista / Testimonio de un huachicolero." *Milenio*. January 18. http://www.milenio.com/policia/robar-a-pemex-es-mas-redituable-que-ser-zeta Date accessed: July 11, 2021.
PEMEX. "Reporte de tomas clandestinas en 2018." https://www.pemex.com/acerca/informes_publicaciones/Paginas/tomas-clandestinas.aspx Date accessed: July 9, 2021.

Pérez Ana Lilia. 2011. *El Cártel Negro: Cómo el crimen organizado se ha apoderado de Pemex*, 5–221. México: Grijalbo.
Pérez Ana Lilia. 2018. "Huachicoleo a escala multimillonaria Dentro de Pemex, toda una industria paralela." *Proceso*. December 29, 2018.
Pérez Caballero Jesús. 2018. "Mexico's CJNG: Local Consolidation, Military Expansion and Vigilante Rhetoric." *Insight Crime*. February 8, 2018. https://www.insightcrime.org/news/analysis/mexico-cjng-local-consolidation-military-expansion-vigilante-rhetoric/ Date accessed: July 14, 2021.
Pérez Esparza David. 2014. "Chapo Guzmán, CEO." *Nexos*. February 23. https://www.nexos.com.mx/?p=19103 Date accessed: July 14, 2021.
Pérez García Julio. 2018. "La Seguridad Energética en los Mercados de Gasolina y Diesel en México." Instituto Español de Estudios Estratégicos. April 5, 2018, 1–24.
Radden Keefe Patrick. 2013. "The Geography of Badness: Mapping the Hubs of the Illicit Global Economy." In *Convergence: Illicit Networks and National Security in the Age of Globalization*, edited by Michael Miklaucic and Jacqueline Brewer, 97–107. National Defense University Press.
Reed Tristan. 2015. "Mexico's Drug War: A New Way to Think About Mexican Organized Crime." Stratfor.
Reinhart, Luke B. 2014. "The Aftermath of Mexico's Fuel-Theft Epidemic: Examining the Texas Black-market and the Conspiracy to Trade in Stolen Condensate." St. Mary's L.J. 45, 750–786.
Reuter Peter. 2014. "Drug Markets and Organized Crime." In *The Oxford Handbook of Organized Crime*, edited by Letizia Paoli, 1–25.
Rodríguez García Arturo. 2019. "Por los Caminos de la Ordeña." *Proceso*. January 27, 2019.
Romero Octavio. 2019a. "Presidential Press Conferenece." Presentation. *Palacio Nacional*. January 14, 2019.
Romero Octavio. 2019b. "Comparecencia del Ingeniero Octavio Romero Oropeza, Director General de Petróleos Mexicanos (PEMEX), ante las Comisiones Unidas de Energía e Infraestructura de la Cámara de Diputados LXIV Legislatura." Presentation. Cámara de Diputados. October 28, 2019.
Sánchez Alejandro. 2012. "Crece 52% robo de combustible; en 2011 detectaron 1,324 tomas ilícitas." *Excélsior*. January 15. https://www.excelsior.com.mx/2012/01/15/nacional/802267 Date accessed: July 14, 2021.
Secretaría de Energía. 2018. "Balance Nacional de Energía 2017," 1–129. Secretaría de Energía.
Solera Claudia. 2017. "Las ordeñas contra Pemex, desde adentro; en lista negra, 130 empleados." *Excélsior*. May 1. https://www.excelsior.com.mx/nacional/2017/05/01/1160791 Date accessed: July 14, 2021.

Stewart Scott. 2018. "Tracking Mexico's Cartels in 2018." *Stratfor*. February 1. https://worldview.stratfor.com/article/tracking-mexicos-cartels-2018 Date accessed: July 12, 2021.

Televisa. 2018. "Quedan Libres Huachicoleros y Uno Sigue Trabajando en Pemex—En Punto Con Denise Maerker." Video. https://www.youtube.com/watch?v=E7mvSS4j-Sk Date accessed: July 14, 2021.

Toledo Guerrero Gabriel. "Corruption in the Mexican Energy Industry: Recommendations and Proposals," 1–25. Wilson Centre Mexico Institute. https://www.wilsoncenter.org/publication/corruption-the-mexican-energy-industry-recommendations-and-proposals Date accessed: July 14, 2021.

Vega Aurora. 2012. "Petróleos Mexicanos ubica 5 mil tomas clandestinas." *Excélsior*. March 19. https://www.excelsior.com.mx/node/819530 Date accessed: July 14, 2021.

Velasco Manuel. Manuel Velasco to Chaipas´ Congress. 2018.

Vergara Rosalía. 2019. "Trabajadores petroleros presentan nueva denuncia contra Romero Deschamps." *Proceso*. February 27. https://www.proceso.com.mx/nacional/2019/2/27/trabajadores-petroleros-presentan-nueva-denuncia-contra-romero-deschamps-220943.html Date accessed: July 14, 2021.

Williams Phil. 2001. "Transnational Criminal Networks." In *The Future of Terror, Crime, and Militancy*, edited by John Arquilla, David Ronfeldt, 61–97. RAND Corporation.

Yashar Deborah J. 2018. "*Homicidal Ecologies: Illicit Economies and Complicit States in Latin America*, 1–368. Cambridge University Press.

PART IV

Guanajuato and the Dispute for Salamanca

A camera mounted on a helmet of one of the perpetrators records everything. Five *sicarios* are riding on two pickup trucks. The vehicles advance a few meters before stopping at a run-down tire-repair shop. A few seconds before coming to a halt, they start shooting at the shop. The camera hitman empties the entire cartridge of his weapon in a few seconds. As he exits the truck, we can see four men accompanying him, wearing ski masks, bullet proof vests and wielding automatic rifles. "All down, get in!" screams the go-pro *sicario*. He advances with one of his men following him. The rest stay covering from the rear-guard. As they advance into the shop, they find a body laying immobile on the floor. The camera-wielding *sicario* shoots him in his forehead without hesitation. They quickly examine the rest of the place. "Anything here?" asks the cameraman. "Nothing" answers his accomplice. Once they confirm that there is nobody there, they return to their vehicles with the video ending abruptly (*Sicarios Kill 5 In Guanajuato* 2019).

Five men were murdered in that shooting. The tire-repair shop was near pipelines and 23 kilometres away from the refinery in Salamanca, Guanajuato. The perpetrators were the Santa Rosa de Lima Cartel (SRLC), a criminal network specialized in fuel trafficking. The video was a part of a strategy to intimidate authorities and rival groups, especially the CJNG. Control over the Salamanca refinery and its pipelines passing through seventeen municipalities (Pérez 2018d) have turned Guanajuato into the epicentre of a violent criminal dispute. Guanajuato concentrates 15% of fuel-theft in Mexico (Dávila 2019) and has positioned itself as a pipeline extraction hotspot:

Reference

"Sicarios Kill 5 In Guanajuato". 2019. Video.

CHAPTER 5

The Local Dynamics of Fuel Trafficking in Puebla and Guanajuato

INTRODUCTION

The consolidation of the MFBM must be understood within Mexicos larger security crisis and, as with the homicide decrease at the beginning of EPNs presidency, local dynamics are crucial for understanding nationwide trends. Fuel trafficking has taken Mexico's criminal networks' presence beyond historic areas of operation and pushed their territorial expansion towards the country's central regions. Interestingly, these enclaves are development hubs with thriving industries and trade connectivity, displaying growth rates surpassing the historically dynamic northern region (Banco de México 2018, 3). The MFBM presents a *"geographic concentration of violence"* (Yashar 2018, 78) in areas of extraction, where energy installations are situated. Competition over extraction, and trade of stolen fuels, had an impact in the violence observed in Mexico during the second half of the 2010s. Groups have the incentives to exercise control over these enclaves to hinder competition, influence or avoid law enforcement, gain market share, and set prices. In this context, Puebla and Guanajuato emerged as MFBM hotspots.

An important factor that influences MFBM operations in subnational territories are supply chains and their geography. Illicit groups must find areas to obtain derivatives and locate destinations for commercialization. Consequently, there is a geographic incentive for criminals to compete for territories considering extraction options and market proximity. This

© The Author(s), under exclusive license to Springer Nature Switzerland AG 2025
S. León Sáez, *Mexico's Fuel Trafficking Phenomenon*, St Antony's Series, https://doi.org/10.1007/978-3-031-70503-8_5

section will highlight the interconnection between illicit flows, criminal conflict and geographic conditions. According to de Boer and Bosetti (2017, 4) territories exhibiting trafficking and criminal conflict tend to have at least one of the following features: (1) are areas with frail governance or large territories with limited state presence; (2) are located along important trafficking routes or are the source of illegal supplies; and (3) present political divisions, corruption and weak governance where determined groups are systematically excluded. The cases of Puebla and Guanajuato will prove to have many of these characteristics.

This chapter will explore the local dynamics of Mexican fuel trafficking, mostly during the growth period between 2011 and 2018. This will be valuable to elucidate phenomena not apparent in national-level analysis, such as socio-economic circumstances, the influence of geographic conditions and the role of local grey actors. They will also help to further explore this research's theoretical framework and analytical tools. We cannot comprehend the Mexican security situation while neglecting local dynamics.

Vignette Four: Guanajuato and the Dispute for Salamanca

A camera mounted on a helmet of one of the perpetrators records everything. Five *sicarios* are riding on two pickup trucks. The vehicles advance a few meters before stopping at a run-down tire-repair shop. A few seconds before coming to a halt, they start shooting at the shop. The camera hitman empties the entire cartridge of his weapon in a few seconds. As he exits the truck, we can see four men accompanying him, wearing ski masks, bullet proof vests and wielding automatic rifles. "All down, get in!" screams the go-pro *sicario*. He advances with one of his men following him. The rest stay covering from the rear-guard. As they advance into the shop, they find a body laying immobile on the floor. The camera-wielding *sicario* shoots him in his forehead without hesitation. They quickly examine the rest of the place. "Anything here?" asks the cameraman. "Nothing" answers his accomplice. Once they confirm that there is nobody there, they return to their vehicles with the video ending abruptly (*Sicarios Kill 5 In Guanajuato* 2019).

Five men were murdered in that shooting. The tire-repair shop was near pipelines and 23 kilometres away from the refinery in Salamanca, Guanajuato. The perpetrators were the Santa Rosa de Lima Cartel

(SRLC), a criminal network specialized in fuel trafficking. The video was a part of a strategy to intimidate authorities and rival groups, especially the CJNG. Control over the Salamanca refinery and its pipelines passing through seventeen municipalities (Pérez 2018b) have turned Guanajuato into the epicentre of a violent criminal dispute. Guanajuato concentrates 15% of fuel-theft in Mexico (Dávila 2019) and has positioned itself as a pipeline extraction hotspot (Fig. 5.1).

The refinery is a focal point for the local MFBM linked to polyducts coming from Tula, Hidalgo and that distribute to Jalisco, Michoacán, the Centre-Bajío and Centre-northern regions. Other local installations targeted by criminals are four Storage and Distribution Terminals (TARs) (Map 5.1).

The presence of this energy installations is not incidental. Guanajuato has developed an important industrial sector attracting investment and increasing energy demand. The state's economy has been growing above Mexicos national average for eight years (Flores 2019) and concentrates 11% of all the nation's industrial parks (Reyes 2018).

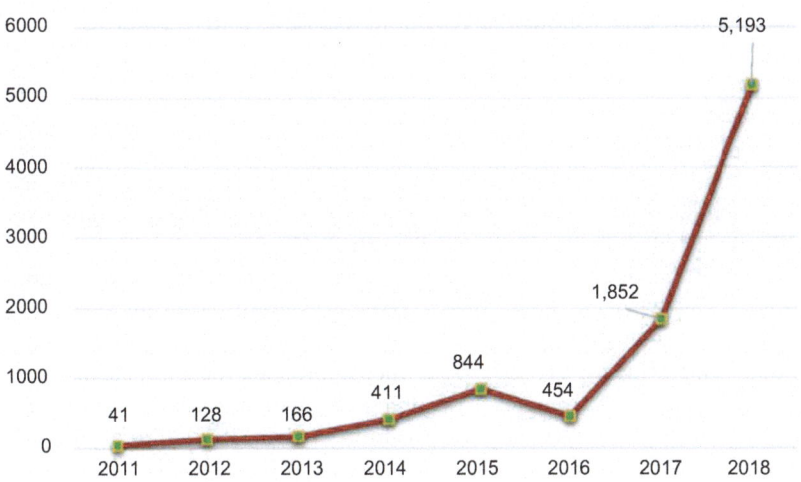

Fig. 5.1 Illegal Extraction Points in Guanajuato (2012–2018) (Transparency Petition Response 1857200119616; *Observatorio Nacional Ciudadano* 2018a, 7)

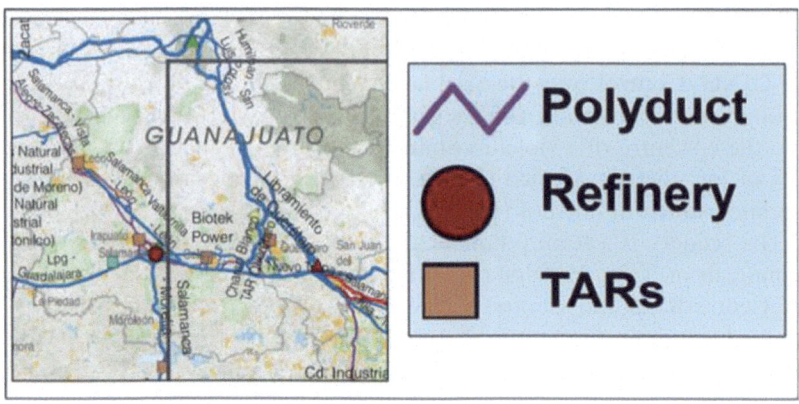

Map 5.1 Energy Installations in Guanajuato (Llano and Flores 2017)

Energy Installations and Criminal Violence in Guanajuato

The Salamanca Refinery is pivotal for homicidal violence in Guanajuato. In the state's industrial corridor, spanning from Celaya to León, homicides increased 225% between 2015 and 2018. As in Nigeria, in Guanajuato there is a connection *"between the oil industry, crime and conflict"* (Jesperson 2017, 4). Guanajuato's criminal violence and economic dynamism is an example of how economic growth generates opportunities for criminal networks. It also indicates the importance of Guanajuato as a relevant energy node with strategic installations facilitating fuel-theft (de Boer and Bosetti 2017, 4).

In 2017, Guanajuato was the state with the most fuel extraction points with 1,852. By 2018, 1,547 illegal extraction points were discovered in the state's pipelines, putting it in third place nationally behind Puebla (2,072) and Hidalgo (1,726). The microregion of the municipalities of Silao, Irapuato and Salamanca occupied the national third place in the number of illegal extraction points found between 2009 and 2016, with a total of 1,399 (EnergeA and Grupo Atalaya 2017, 40).

Between 2009 and 2016, Irapuato was the second municipality with the highest number of illegal extraction points detected nationally, with 731. This data points to the correlation between the dispute over the local MFBM among criminal networks and growing violence. This connection has been made by authorities: in December 2018, Guanajuato's governor declared that 85% of violent homicides in the state were linked to fuel trafficking (de Mauleón 2019). According to the think-tank *México Evalúa*, homicidal violence in Guanajuato is concentrated in municipalities where fuel-theft increased explosively in recent years (México Evalúa 2018). Academic research has found a similar correlation between increasing criminal violence and the territorial competition for subnational enclaves that concentrate lucrative black markets. Yashar (2018) argues that territorial control is an essential requirement for criminal actors to *"sustain their livelihood and turn a profit"* (72). This control is achieved through violent competition leading to expanding homicidal violence.

Guanajuato is an example of a valued criminal subnational enclave. For illicit markets and criminal groups, in Mexico and other criminally violent Latin American countries, *"geographically, violence-prone subnational enclaves"* where *"security forces are weak and/or corrupt"* (Yashar 2018, 4) are emerging as the most priced assets. Having explained this, we will now delve into the groups responsible for making Guanajuato a violent fuel trafficking hotspot.

The Criminal Networks Behind the MFBM in Guanajuato

Guanajuato has different criminal groups trying to profit from the MFBM, from large networks to criminal gangs, with specialized mid-size fuel trafficking groups in-between. Official sources point to the presence of criminal networks from Michoacán such as the Knights of Templar (EnergeA and Grupo Atalaya 2017, 110). Criminal cells of the GC and the Zetas have historically had presence in Guanajuato (Saucedo 2019). Another group is the SRLC, whose origins are connected to a fragmentation from a larger network (the CJNG) and to an alliance among local networks from the Tierra Caliente region[1] to confront the CJNG's expansion (Stargardter 2018a; Reina 2019; Dalby 2019).

The SRLC operates as a Specialized Fuel Trafficking Group: using professional equipment, employing sophisticated fuel-theft methods (removable hermetic pipeline valves), and engaging in refinery theft

---

[1] The Tierra Caliente region occupies the southwest Pacific coast (Jalisco, Manzanillo, Michoacán, Guerrero) and stretches inland into the Bajío and Centre areas of Mexico (Guanajuato, Querétaro, Hidalgo, Estado de México and Puebla). This region has spawned groups such as the Knights Templar, Cártel de Jalisco Nueva Generación, La Familia Michoacana and Guerreros Unidos.

co-opting workers in Salamanca and local authorities. They have military-grade weaponry, manage large storage sites, maintain strongholds in the south of Guanajuato and have expanded into Querétaro and Hidalgo. It is estimated that the SRLC has the capacity to steal 6,100 barrels of hydrocarbons daily (Dávila 2019), making this enterprise its most important source of profits and its criminal cells have been exploiting it for at least 12 years (Saucedo 2019). Other smaller networks are also striving to profit from the Guanajuato MFBM. One example is a group known as *Los Pelones* that steal 2,500 barrels of hydrocarbons daily (Dávila 2019).

Different sources point to the confrontation between the CJNG and the SRLC over Guanajuatos MFBM as one of the main factors behind increasing violence (Stargardter 2018a; Mohar 2018). The CJNG appeared in Guanajuato by forming sub-networks with local criminals, implementing a franchising system (Stargardter 2018a). This observation, made by a witness of the CJNG's operations in Salamanca, agrees to what was stated by Jones (2018) who argues that this group had been annexing and licencing criminal cells remnants of fragmentation processes in Mexico's illicit underworld since 2015 (20). Crosschecking with local information sources also corroborated this *modus operandi* in which the CJNG, during its first incursions into Guanajuato, would send small armed groups to recruit and train locals to carry out their criminal activities within the state (Saucedo 2019).

The CJNG still must be researched, but evidence indicates that it is a highly resilient and geographically dispersed criminal network that operates with the group of *Los Cuinis*, a powerful financial and money-laundering wing. This partnership has made the CJNG the most economically powerful Mexican criminal network (DEA 2018) with the fastest access to large amounts of cash (de Mauleón 2015). The licencing expansion model seems to be part of the CJNG's DNA. Apart from the influence of the Zetas' expansion model (Garay-Salamanca and Salcedo-Albarán 2016, 87), leadership figures of the CJNG (such as Nemesio Oseguera *El Mencho*) were once part of the Milenio Cartel, an offshoot group of the Sinaloa Cartel. This indicates that the CJNG leadership is familiarized with the licencing expansion model. The CJNG assimilated fragmented criminal cells as the security crisis escalated in the Tierra Caliente region. This strategy boosted the territorial expansion of the CJNG (de Mauleón 2015). In 2014 security sources declared that the CJNG backed self-defence groups to wage a proxy-war against the Knights of Templar in Michoacán ("El Abuelo, de autodefensa..." 2018).

The CJNG expansion is reinforced by the advantages offered by the Tierra Caliente region (which includes their Jalisco stronghold), like engaging in methamphetamine and opiates transnational trafficking enabled by the Pacific ports of Manzanillo and Lázaro Cárdenas (Stewart 2018). Control over valued criminal assets has given the CJNG access to vast resources to sustain its nationwide expansion. Adding to this, the CJNG has shown to be a versatile criminal network capable of opening fronts of confrontation in certain areas while pursuing local agreements in others capitalizing on fragmentation (Jones 2018, 33).

In the northwest and the central-west areas of Guanajuato, the CJNG has important strongholds (Saucedo 2019). These areas concentrate the municipalities of León and Silao which form part of the thriving industrial corridor. Within this corridor, the municipalities of the centre of the state are bastions of the SRLC, including the municipality of Salamanca where the refinery is located. The municipalities of the industrial corridor are experiencing a criminal conflict between the CJNG and the SRLC. There are also confrontations between criminal cells from Michoacán and the CJNG in southern areas of the state (Map 5.2).

This dispute between the SRLC and the CJNG seems to have developed into a proxy war confrontation just as it did in Michoacán. It appears that the GC, the Zetas and the Sinaloa Cartel are supporting the SRLC and deployed armed cells under the brand *Cárteles Unidos* that are confronting the CJNG in Guanajuato and the Tierra Caliente region. What these groups are pursuing with these actions is containing the expansion of the CJNG, disputing this key area for Mexico's criminal underworld and protecting their northern strongholds from an encroachment by the powerful *Jalisciense* group. Local analysis points to the decisive support that an armed wing of the GC (*Grupo Sombra*) gave to the SRLC that led to an impasse of continuous criminal confrontation in Guanajuato (Saucedo 2019).

These groups operate co-opting PEMEX insiders, especially in the Salamanca refinery. As one informant of the SRLC said, the refinery was *"the brain"* of their operation (Stargardter 2018b). Here it is important to explain the co-option dynamics of refinery workers. Many of them are threatened to coerce them into supplying access, technical expertise or valuable information. Workers who do not comply are left defenceless by institutions incapable of protecting them. My source who covered the situation in Salamanca documented the case of a refinery pump technician who was kidnapped and stabbed by members of the Familia Michoacana

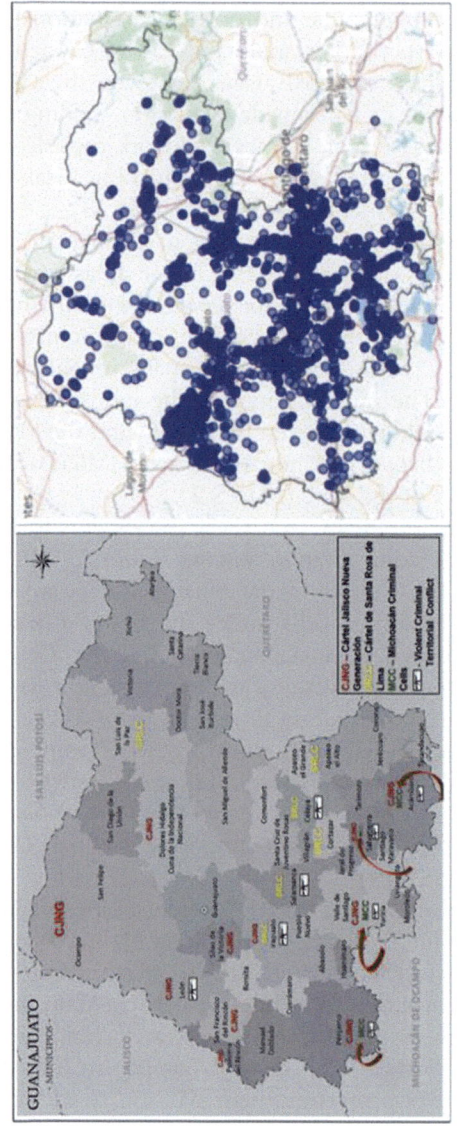

Map 5.2 Criminal conflict and criminal incidence in Guanajuato (Elaborated by author with information of Saucedo 2019)

crime network after declining to give them information on the hours that fuel was dispatched from the refinery to the pipelines. He looked for help within PEMEX and was ignored. He subsequently had to leave Mexico applying for refugee status in 2016.

Because of this institutional neglect and impunity various PEMEX workers lost their lives. Local journalism points to the murder of at least seven PEMEX workers between 2012 and 2018 (Stargardter 2018b). One of the most publicized cases was the murder of the head of Physical Security of PEMEX in Salamanca who oversaw the installations in the refinery and Guanajuato's fuel distribution network ("Asesinan al jefe de Seguridad Física..." 2018a). Co-opting the refinery allows to capture vast amounts of revenues for criminals, useful to co-opt local representatives and law enforcers. These are two grey actors still to be developed who play significant roles in local fuel trafficking.

Guanajuato: Co-opting Local Authorities

Co-option of municipal police forces in Guanajuato is widespread. In Salamanca, the local police force disbanded after the kidnapping and homicide of three of its agents in August 2017. This took place after the abducted policemen claimed on a video that the police force was fully co-opted by criminals. After this incident, agents stopped showing up for their duties altogether, leaving the municipality without a police force (Enfoque 2017). State and municipal police forces in Guanajuato have been co-opted by the CJNG and the SRLC, giving them access to armed cells that include vehicles and communications equipment. Their involvement within both criminal networks (achieved through bribes or coercion, or a combination of both) has been such that some police forces have made a transition from playing *"partially logistical roles"* related to cargo to behaving as armed wings (Saucedo 2019). In 2019, Guanajuato was the state with the highest number of murdered policemen and had the highest national desertion rates (Alto Nivel 2019).

The SRLC built a support network including politicians and police. These grey actors have proven to be an invaluable asset for this group. This network includes mayors and police forces (municipal police directors, deputy prosecutors and commanders of the state security forces) (Saucedo 2019). All these agents have provided resources (from information to financial support), protection and impunity for this network and the many black markets it exploits. With these varied number of grey actors, the SRLC has proved to be an extraordinarily resilient criminal network with the resources and capacity to engage in open confrontation against the state and macro-criminal networks simultaneously. Within this diversified grey actor network there are powerful agents with the capacity to extend the overall scope of entire criminal operations (Williams 2001, 81).

A leaked call between a member of SRLC and Hugo Estefania Monroy, former mayor of Cortazar and member of the state leadership of the Democratic Revolution Party (PRD), revealed their intentions to illegally obtain federal resources through municipal governments, buy weaponry and extort local businesses (""Quieren ordeñar apoyos..." 2019; Noticieros en Línea 2019). After that call was made public, Guanajuato's Attorney General announced Estefania Monroy was being investigated. On November 30, 2019, Estefania Monroy was murdered by an armed commando in the municipality of Cortazar ("Comando asesina a Hugo", 2019).

That a high-ranking political grey actor is involved in the SRLC is telling. It reveals an active interest from different actors to sustain criminal conflict. According to Jesperson (2017), illicit activities linked to conflict can provide cover for criminality and create *"a shared strategic space"* for illegal groups to increase revenues. This generates an appeal for multiple actors to maintain armed conflict, continuously generating profits to sustain violent confrontations (7).

THE PUEBLA CASE: OF DEVELOPMENT AND CRIMINAL GEOGRAPHIES

Puebla is a fuel-theft hotspot where the Zetas, the CJNG, specialized groups and community gangs are involved in exploiting different pipelines, mostly the ones located in the east of the state towards Veracruz and in the west towards Mexico City. As a result, illegal extraction points in that state have shown consistent growth (Fig. 5.2).

One piece of data shows Puebla's relevance in the MFBM: with a pipeline network of 131 kilometres, it concentrated over 22% of illegal extraction points nationwide in 2016 (1,533 out of 6,873). In 2018 Puebla was the state with the highest number of syphon points nationally. The MFBM in Puebla is also unique for the large-scale involvement of its local communities. Puebla also illustrates how geographic, economic,

Fig. 5.2 Illegal extraction points in Puebla (2011–2018) (Gobierno Fácil 2019; PEMEX: "Reporte de tomas clandestinas")

social and security phenomena combine to create one of the most thought-provoking examples of Mexican fuel trafficking.

Community Participation in the Poblano MFBM

Many municipalities involved in fuel trafficking in Puebla are agrarian communities that have suffered exclusion and poverty since the 19th century (Mastretta 2018). These communities have been historically omitted from development projects since then or, in other cases, inclusion led to exploitation. Likewise, Puebla has undergone structural changes in the last 40–50 years. During that period, a system of highways connecting Puebla to Veracruz and to Mexico City was developed, as well as 11 fuel pipelines (mostly polyducts). Large industry projects were built, like the Volkswagen plant and the industrialization of the northern area of the state capital. This created development poles that reconfigured Puebla to its core.

Along the way, many of these excluded agrarian communities (impoverished by transgenerational segmentation of their small plots of land) found themselves surrounded by different development hubs. This combination of exclusion, development and strategic geographical location set the stage for fuel trafficking in Puebla. Community fuel-theft started organizing before the involvement of seasoned criminals. Many community-based fuel traffickers did not have permanent jobs, and many migrated. Suddenly, through contacts in neighbouring towns, they began to organize groups dedicated to fuel-theft through rudimentary pipeline extraction. Subsequently, entire families found sustenance in this illicit activity, with leaders becoming local benefactors.

Exclusion from development projects was reinforced by political decision-making of Governor Moreno Valle (2011–2017) of the National Action Party (PAN). Amongst the seventeen municipalities concentrating the largest number of extraction points between 2009 and 2016, four were in Puebla: Acajete, Acatzingo, Palmar de Bravo, Quecholac and Tepeaca. These municipalities concentrated 70% of the state's illegal extraction points during that period (EnergeA and Grupo Atalaya 2017, 102), an area subsequently known as "The Red Triangle". In these areas 80% of the population are poor (Enciso 2017). Under Moreno Valle, patronage programmes supporting impoverished agrarian communities were eliminated. This was part of an *"aggressive and successful"* project to dismantle political networks that the Institutional Revolutionary Party (PRI) historically managed in Puebla (Mastretta 2017).

Journalist Edmundo Velázquez (2018) explained that the elimination of patronage programmes affected the municipalities of Acajete, Acatzingo, Tepeaca, Palmar de Bravo and Palmarito in particular, because of their reliance on vegetable production. This miscalculation allowed criminal groups to gain support by enabling sources of income through fuel trafficking. In those towns people employed in washing produce were paid from $50 to $100 MXN daily ($2.6 USD–$5.2 USD), while fuel traffickers paid from $1,000 to $1,500 MXN ($53 USD–$79 USD) for driving a truck of stolen hydrocarbons for a few hours during night-time (Ibid). From there, incidents took place in which communities confronted authorities seizing stolen fuels and others in which entire towns collected fuel from a punctured pipeline while authorities stood as witnesses ("¿Qué Pasó el 3 de mayo…" 2017).

Other political decisions affected Puebla communities pushing them into fuel trafficking. In the state's Sierra Norte region the extinction by presidential decree of *Luz y Fuerza del Centro*, a public monopoly supplying electricity service in the centre of Mexico, in 2009 left 40,000 families without income sources (Velázquez 2018). This created another catalyst for locals to get involved in the MFBM. These two events contributed to create areas where local populations were susceptible to criminal co-option, one in the north and another in the centre of the state, both concentrating pipelines (Map 5.3).

This connection between criminals and communities is recognized by official sources. In the town of Palmarito there is involvement of the local population in pipeline extraction, storage and smuggling of stolen fuels. Government sources sustain that Zetas criminal cells operate in Palmarito, where the population are *"highly reactive"* to authority intervention and confront *"elements of the federal forces to defend their activities"* (EnergeA and Grupo Atalaya 2017, 99). Locals justify their involvement and their criminal accomplices saying job opportunities are scarce, poorly paid and that their participation is due to fuel price increases. Both journalists that contributed for this research corroborated these assertions (Mastretta

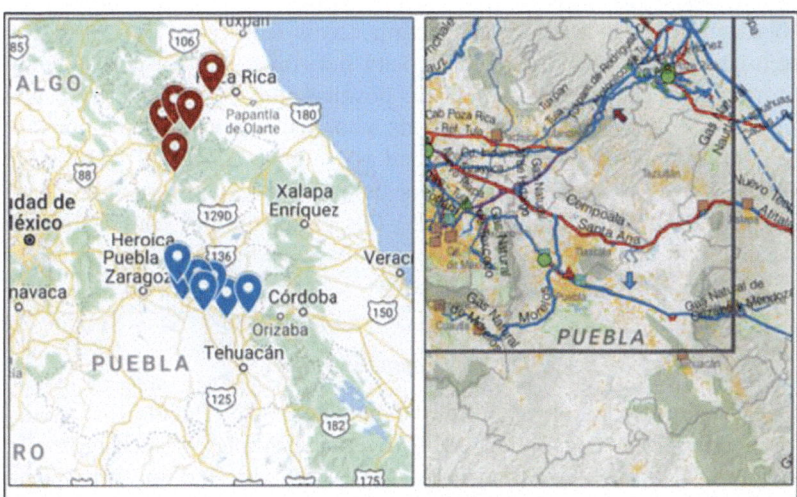

On the map: the municipalities of the Red Triangle region in blue and the municipalities affected by the extinction of *Luz y Fuerza del Centro* in 2009 in red. In the northern region of Puebla runs the Tuxpan – Tula pipelines connecting Veracruz to the Tula Refinery in Hidalgo (red and blue arrows) (Llano and Flores 2017).

Map 5.3 Municipal fuel trafficking hotspots and energy installations in Puebla

2018; Velázquez 2018) and local *corridos* and *cumbias* mention fuel trafficking as a response to fuel price increases (Los Reyes Klan 2017; Kary Siza y Los Chikos Kumbia 2017).

Illegal enterprises can have negative consequences for the communities where they develop (violence, resource predation, addiction, human rights violations). But in some circumstances, black markets can generate benefits. In such cases illegal enterprises can mutate into socially accepted activities. Academics have found illegal markets that generate expectations of an improved future amongst their participants. As Beckert and Dewey (2017) argue, illegal markets *can "become economic structures that provide access to at least a minimum level of economic citizenship for some"* and that these unlawful enterprises *"need to be analysed in close connection with (the lack of) available alternatives and the aspirations of the populations carrying out these activities"* (15).

Puebla's community involvement in fuel smuggling amongst local populations, who have been subject to historic exclusion, shows how a black-market can expand by offering access to income opportunities, essential goods and services, reciprocity networks, and by guaranteeing autonomy for its participants. These positive effects can lead to a legitimization of illegal market participants who challenge the unlawfulness of their actions, a phenomenon described as *"contested illegality"* (Hübschle 2017). These are strong motivators behind the expansion of Puebla's fuel trafficking. And as shall explore, this exclusion and contestation of illegality combine with development hubs allowing fuel smuggling to thrive.

The Development Hubs

Development created economically dynamic hubs in Puebla that would play decisive roles in the state MFBM. Four examples of such hubs are: the municipality of Cuautlancingo, the market located in San Martín Texmelucan, the area north of the state's capital and the supply central of Huixcolotla.

Cuautlancingo's Volkswagen plant is the largest in North America and is responsible for 78,000 jobs (Martínez 2015). Texmelucan has the largest textiles market in Mexico, attracting 50,000 visitors weekly. The economic resources of this market are enormous: it is a sales-point for contraband and counterfeits, activities that attract criminal actors. Texmelucan also has large industries producing petrochemicals, auto parts and textiles. Moreover, a large agro-industrial sector has developed in the north of Puebla. Finally, in Huixcolotla, we find the second largest supply centre in the country where 3,000 tons of food are commercialized daily. This generates great demand for freight transport consuming stolen fuels in bulk. The centre even has its own area dedicated to selling stolen derivatives (Mastretta 2018; Velázquez 2018; France 24 2017). The development of dynamic economic hubs allured criminal networks into Puebla and created demand for the MFBM.

Since the mid-2000s the Zetas start expanding their criminal networks from Veracruz to Puebla. The Beltrán Leyva network (an offshoot of the Sinaloa Cartel) soon followed. These networks were pursuing the exploitation of other black markets linked to economic development: kidnapping, extortion, drug dealing, piracy, prostitution. Still, their presence in the state was not as profound, nor was their involvement in the MFBM. One project would change that.

The project detonating large-scale fuel trafficking in Puebla was the construction of a plant of the automobile manufacturer Audi. For this project, the Moreno Valle administration constructed a platform for Audi's investment. This operation required clearing a swath of land for construction and approximately 2,000 trucks working 24/7 removing rubble between 2012 and 2014. And yet again another grey actor played a crucial role in this case. The Confederation of Mexican Workers (CTM) in Puebla was in charge this clearing operation. The leader of the CTM in the state at the time (and congressman with links to Moreno Valle), Leobardo Soto Martínez, ran the clearing operation acquiring stolen diesel. This event would revolutionize the MFBM in Puebla.

As Mastretta and Silva (2019) argue, *"the [fuel-theft] explosion in Puebla has its start at that juncture of 2013–2014, when the transporters controlled by the CTM obtained the fuel at 8 pesos a litre"* (95). The trucks used for this operation made 10–14 daily trips requiring between 80–100 litres of diesel. The scope of this operation was such that fuel traffickers in Tepeaca and Acatzingo offered CTM truck drivers stolen diesel at a preferential price of $4 MXN/litre ($0.21 USD) (Ibid.). The profits this operation generated made fuel trafficking in Puebla an attractive enterprise for criminal macro-networks like the Zetas and, subsequently, the CJNG (Mastretta and Silva 2019).

With the construction of the Audi plant, an MFBM triangle consolidated in Puebla with four consuming hubs (Map 5.4).

Map 5.4 MFBM consuming hubs in Puebla (Elaborated by author with information of Mastretta 2018)

All these projects made Puebla a place with industrial-scale consumers for stolen fuel, explaining part of the growth in fuel extraction points between 2012 and 2014 that exploded in 2015. Affluent local actors are better-off acquiring stolen fuel to sustain their operations, at prices below those of the legal market which had increased considerably. In this context diesel, mostly used in industry and freight transport, was the most stolen fuel in Puebla. Development created consuming hubs for the poblano MFBM, materializing the means for large-scale operations. For example, pipelines in Puebla coming from Veracruz run parallel to the 150D highway giving efficient access to fuel tanker trucks. Puebla's highway network makes it a state with valuable land mobility. It is two hours away from Mexico City and Veracruz, and four hours away from Tamaulipas. Puebla neighbours Hidalgo, another fuel trafficking hotspot. This makes Puebla a connectivity node opening multiple destinations for stolen fuel, and criminal networks exploit it accordingly (Velázquez 2018).

Criminal Networks Behind the MFBM in Puebla

Puebla's location and economic development created profit opportunities for criminal networks. The first signs of the presence of the Zetas in the state began between 2006 and 2008 in northern municipalities bordering Veracruz (Ibid). This group started using co-opted municipal police agents to obtain extortion revenues in the municipality of Esperanza and controlling illegal businesses. This municipality acted as a gateway for the group to access Puebla's "Red Triangle" ("Cómo operan los huachicoleros..." 2017a).

One of the first protagonists of the criminal incursion into Puebla is Roberto de los Santos de Jesús, *El Bukanas*. He was a municipal policeman from Veracruz, who began his criminal career extorting businesses in Puebla. Subsequently, *El Bukanas* became the manager of a network specialized in fuel trafficking linked to the Zetas ("¿Quién es el Bukanas?" 2018). With their incursion, the Zetas brought their fuel trafficking experience. During the Moreno Valle administration, it was estimated they controlled 85% of Puebla's MFBM (EnergeA and Grupo Atalaya 2017, 102) through a sub-network named *Sangre Nueva Zeta*. Criminal subsidiaries started appearing in Puebla as the Zetas consolidated their presence by creating sub-networks and co-opting others, diversifying into extortion, kidnapping and piracy. These new groups started sharing their territory with other criminals who relied on local predatory enterprises.

By 2015, the CJNG started showing signs of its presence in Puebla. An ongoing violent confrontation between the CJNG, the Zetas and their affiliated networks escalated when a leader linked to *El Bukanas* criminal network was assassinated in October 2017. Ever since the irruption of the CJNG, criminal violence in Puebla has grown dramatically. In 2012 Puebla registered 465 homicides. This grew to 568 in 2014, 632 in 2015, 735 in 2016 and a record high of 1,070 in 2017 (INEGI 2018). In seven years, homicides grew by over 130%. Vehicle theft, a practice linked to fuel trafficking, also experienced dramatic increases, going from 698 in 2015 to 1,182 in 2016 and 3,630 in 2017 (SESNSP).

Puebla presents a deeply diversified illicit landscape with large and medium criminal networks, all disputing the MFBM with community-based fuel-theft gangs. Groups like *Los Bukanas* and *Los Téllez* are medium-size criminal networks with links to the Zetas and the CJNG respectively. It is estimated *Los Téllez* have an operational capacity to steal 6,900 barrels of fuel daily, *Los Bukanas* steal approximately the same amount. Groups like this share the state with independent groups. These groups, all Specialized Fuel Trafficking Networks, have diversified into neighbouring states like Estado de México, Tlaxcala and Veracruz.

In central Puebla, local authorities identified fifteen criminal groups operating along the Mexico City–Veracruz highway (Velázquez 2019a). With its diverse criminal landscape, Puebla became a pipeline theft powerhouse: between 2011 and 2017 illegal extraction points in Puebla increased an astounding 1,096.68% (Observatorio Nacional Ciudadano 2018b, 2). By 2018, Puebla was the nationwide leader of illegal extraction points, with 2,072, representing 13% of nationwide pipeline fuel-theft (Dávila 2019). This growth of Puebla's MFBM occurred with grey actor involvement.

The Grey Actors in Puebla's MFBM

The grey actors involved in the MFBM in Puebla are astonishingly diverse. One example is union leader and congressman Soto Martínez, a key figure behind the explosive growth of fuel trafficking between 2012 and 2014. His terrain-clearing operation for the Audi plant done with stolen fuel, explains the expansion of the MFBM in Puebla. Soto Martínez, with his political connections and access to legal and illegal resources, is an exemplary network-extending grey actor (Williams 2001, 81). Another grey actor mentioned previously is *El Bukanas*, who began as a co-opted police agent and later transitioned into a criminal network manager.

Los Téllez criminal network also operate through grey actors. This group has been linked to a former federal congresswoman and have the protection of a local leader of *Antorcha Campesina* (Ibid.), a patronage agrarian movement linked to the PRI. This group has a long history of illegality and are known for their involvement in violent protests, murders, kidnappings and occupation of private property. It was revealed that leaders of *Antorcha* owned fifty gas stations linked to stolen fuel distribution (Cabrera and Velázquez 2019; Loret 2019).

Regarding grey actor involvement in the MFBM in Puebla, the Moreno Valle administration marked a turning point. Under Moreno Valle, large-scale co-option of municipal governments by fuel traffickers took place, and collusion would reach as high as Puebla's state government. In July 2015, the Chief of the State Police Marco Antonio Estrada, was detained escorting a 147-vehicle convoy transporting stolen fuel. The police accompanying Estrada later denounced the state Security Secretary, Facundo Rosas, for receiving $10 million MXN ($499,724 USD) a month from fuel traffickers (Velázquez 2018). Rosas resigned, but no legal process followed. Only Estrada was imprisoned. According to my journalistic sources, the arrival of Rosas as the highest security official in Puebla marked the entrance of law enforcers who were involved in fuel trafficking in other states.

Fuel trafficking networks found powerful grey actors in Puebla's government while a process of widespread co-option occurred at the municipal level. For the elections of 2013, fuel traffickers financed municipal campaigns co-opting local governments (Velázquez 2018). The co-option of municipal governments of all political affiliations was profound. Some cases include:

   I. The mayor of Palmar de Bravo, Pablo Morales Ugalde (2014–2018), who in July 2017 was detained by Marines for fuel trafficking and money-laundering. The politician, whose candidacy was supported by a coalition including parties like PAN and PRD, made unjustified financial operations in 2015 for $57 million MXN ($2.9 million USD) and acquired seventeen properties with illicit resources. Without explanation he was liberated in October 2018.
  II. The former mayor of Quecholac, a member of the PRI named Néstor Camarillo, was investigated by the Federal Attorney General for links to *Los Bukanas* criminal network.
 III. The mayor of Atzitzintla, José Isaías Velázquez, who was detained by the State Attorney General in 2017 for fuel trafficking links.
  IV. Eleven mayors were investigated in 2017 for enabling fuel-theft (Jiménez 2019a).
   V. In 2018, 50,000 litres of stolen fuel were found in a property belonging to the mayor of Venustiano Carranza. After this incident, legislators called for the investigation of the governor because of his links to this mayor (Velázquez 2018).

Municipal co-option pursued control over local police forces. In San Martín Texmelucan, the entire police force was involved with a criminal cell specialized in fuel-theft. It turned out 119 policemen were not registered, and stolen fuel was found in their headquarters. The state government arrested all 186 enforcers in May 2018 ("Desmantelan policía municipal…" 2018). The far-reaching co-option levels fuel trafficking achieved in Puebla make it a distinctive case in which large numbers of municipal governments (and even the governor's office) showed involvement.

The MFBM in Guanajuato and Puebla

Developing these two state-level case studies gives us some important insights regarding fuel trafficking local dynamics. By exploring the cases of Puebla and Guanajuato one of the most enlightening elements is the role local grey actors play in local MFBMs. Co-opting local mayors, state-level officials and security agents guarantee ground-level impunity for fuel-theft to thrive. Local police also serve as lookouts and criminal armed groups.

Mayors can have important roles regarding distribution. As Ana Lilia Pérez argued (2018a), local politicians tend to own gas stations in their municipalities. This makes them ideal actors to co-opt for supply and money-laundering. This exemplifies how criminal networks, operating in subnational territories, depend on grey actors functioning as *"local fixers"* who have investments and provide *"connections to the market and financial networks needed to extract and sell the commodity"* (Farah 2013, 78). State authorities provide a layer of protection against interreference (from federal security forces or criminal competitors) and can facilitate network-expanding opportunities, as in the cases of the Audi plant in Puebla and of Estefania Monroy in Guanajuato (offering criminal actors the possibility of capturing federal resources through municipalities). These network-extending actors give criminal networks opportunities to increase the scale of their enterprises through their political, economic and legal resources.

Both cases show the importance of location for local fuel black markets to develop. Puebla's position between the energy-rich Gulf of Mexico and Mexico City makes it a strategic pipeline node, in addition to its industries generating energy demand. Guanajuato shares some of these characteristics, positioned between the Tula refinery in Hidalgo to the east and the Centre-Bajío and Centre-north regions to the west. Yet this state also includes its own refinery, making it a production and transport hub, while also being a development centre. Both MFBMs are influenced by their geography and economic dynamics. Both cases concur with Garzón (2014) regarding growth and criminality: Guanajuato and Puebla are economically dynamic states whose development made them attractive territories for criminal networks. Fuel trafficking exacerbated this illicit appeal.

Here the criminal hub argument proves relevant: criminal networks require operational centres with *"advantages found in strong states— infrastructure, banking"* (Miklaucic and Brewer 2013, XVII) combined with state degradation, weak state presence, corruption, poverty and informality. Mexico, mixing developed hubs like Puebla and Guanajuato with state limitations, exclusion and poverty, has become an ideal environment for criminal networks.

The levels of community involvement in Puebla and the contested illegality sustaining it are distinctive of the poblano MFBM and are rooted in historical phenomena. Community backing is complemented by the co-option of local representatives permeating multiple instances of the local state apparatus: from mayors to the governor, from municipal police forces to the state police and the Security Secretariat. Puebla and Guanajuato show the structural deficiencies afflicting state and municipal police forces across Mexico making their agents highly vulnerable to co-option and criminal violence.

Police forces are the institutions in charge of monitoring and regulating public spaces and are the state's first line of contact with criminal groups (Yashar 2018, 105–106). This makes these entities the institutional first layer when we refer to state capacity regarding public security and the rule of law. Therefore, members of the police, along with actors of the criminal justice system, are critical for criminal organizations determining their subnational areas of operation (Ibid., 100–101). Weak and underdeveloped local police forces are more likely to be co-opted creating a protective layer for criminal operations. Both of our local cases substantiate this argument. Leaderships of labour and social organizations (CTM and *Antorcha Campesina*) also got involved and played roles in Puebla's MFBM, contributing to its unprecedented expansion. This also leaves a lesson about economic development projects plagued by exclusion, creating opportunities for profitable criminal enterprises to build social support bases.

Guanajuato presents a phenomenon of criminal network conflict over control of the local MFBM linked to the state belonging to the Tierra Caliente region (Reed 2015). This is reinforced by the involvement of groups like the Knights of Templar that originated in the neighbouring state of Michoacán, and the CJNG from the state of Jalisco. The Specialized Fuel Trafficking Group SRLC itself is the result of a secession from the CJNG and criminal groups from the Tierra Caliente. The key players in this subnational fuel trafficking enclave are all somehow connected to this highly disputed criminal region. It is important to reiterate that the groups involved in these territorial conflicts can vary in capacity and, transnational macro-criminal networks participate in local and regional black markets deploying a cell-based structure. These enterprises can generate profits complementing (or making up for the loss of) revenues of transnational narcotics. The same could be said about Puebla, where the Zetas incursion is linked to its location neighbouring Veracruz, but in Guanajuato's case it helps further understand its characteristic criminal conflict over the state's MFBM.

The irruption of the MFBM in Mexico's volatile criminal landscape displaced violence into new geographies, and criminal groups have responded to crackdowns showing resilience and adaptability. A crossfire in 2017 in Palmar de Bravo with military forces led a group linked to the CJNG to displace their operations from the east of Puebla to the west (Mastretta 2018), exemplifying the mobility of these groups. The Obrador crackdown in Puebla led to a growth in other criminal activities like freight transport theft, theft of cargo trains and kidnappings (Rivas 2019). In Guanajuato, the federal government aggressively pursued the involvement of the SRLC in the local fuel black market. In response, the SRLC *"strengthened its "division" of kidnapping and extortion and has ventured into the theft of ATMs and banks"* (Saucedo 2019).

In May 2019, journalistic sources pointed out that fuel traffickers in Puebla had, after state enforcement presence increased in the Red Triangle, changed their operations to the northern municipalities of the state taking advantage of the absence of federal forces in those areas (Velázquez 2019b). These responses attest to criminal networks' adaptability in relation to fuel trafficking and point to the limitations of the state when trying to crackdown on criminal enterprises across Mexico's vast territory.

All these dynamics show the importance of contemplating local contexts while analysing security matters. Each case shows us similarities and differences essential to achieve a deeper understanding of the phenomena combining from local contexts, shaping the deeply complex fuel trafficking situation unfolding in Mexico at the national level.

Bibliography

"¿Qué Pasó el 3 de mayo en Palmarito, Puebla?" *Noticieros Televisa*. May 11, 2017. https://noticieros.televisa.com/ultimas-noticias/que-paso-3-mayo-palmarito-puebla/.

"Asesinan al jefe de Seguridad Física de Pemex en Salamanca." *Noticieros Televisa*, January 25, 2018a. https://noticieros.televisa.com/ultimas-noticias/asesinan-jefe-seguridad-fisica-pemex-salamanca/. Date accessed: July 12, 2021.

"Comando asesina a Hugo Estefanía Monroy, ex alcalde de Cortazar Guanajuato." *Denise Maerker*, December 2, 2019. https://www.facebook.com/den isemaerkeroficial/videos/comando-asesina-a-hugo-estefan%C3%ADa-monroy-ex-alcalde-de-cortazar-guanajuato/437647897128336/. Date accessed: July 13, 2021.

"Cómo operan los huachicoleros del Triángulo Rojo, un reportaje de Carlos Loret." *Noticieros Televisa*. May 18, 2017a. https://noticieros.televisa.com/ultimas-noticias/como-operan-huachicoleros-triangulo-rojo-reportaje-carlos-loret/. Date accessed: July 13, 2021.

"Crisis en Policía de Salamanca; refuerzan con agentes de FSPE." *Enfoque*. August 22, 2017. http://elotroenfoque.mx/crisis-en-policia-de-salamanca-ref uerzan-con-agentes-de-fspe/. Date accessed: July 12, 2021.

"Desmantelan policía municipal en San Martín Texmelucan." *Excélsior TV*. May 2, 2018.

"El Abuelo, de autodefensa a operador del CJNG." *El Universal*, May 30, 2018. https://www.eluniversal.com.mx/nacion/seguridad/el-abuelo-de-autodefensa-operador-del-cjng. Date accessed: July 12, 2021.

"Estos son los estados con más homicidios y los policías mejor y peor pagados en 2019." 2019. *Alto Nivel*, December 17, 2019. https://www.altonivel. com.mx/actualidad/estos-son-los-estados-con-mas-homicidios-y-los-policias-mejor-y-peor-pagados-en-2019/. Date accessed: July 13, 2021.

"Justice, Treasury, and State Departments Announce Coordinated Enforcement Efforts against Cartel Jalisco Nueva Generación." *Drug Enforcement Administration (DEA)*, October 16, 2018. https://www.dea.gov/press-releases/2018/10/16/justice-treasury-and-state-departments-announce-coordinated-enforcement. Date accessed: July 12, 2021.

"Quieren ordeñar apoyos federales." *Reforma*. March 13, 2019.

"Sicarios Kill 5 In Guanajuato." 2019. Video.

Ana Lilia Pérez (Independent Journalist and Author) Interviewed by Samuel León in Mexico City, August 23, 2018a.

Banco de México. 2018. "Reporte sobre las Economías Regionales." Banco de México, pp. 1–40.

Cabrera Yonabad and Velázquez Edmundo. 2019. "Diez líderes de Antorcha Campesina, entre los propietarios de gasolineras en Puebla y Guerrero," February 18. https://www.periodicocentral.mx/2019/pagina-negra/huachi col/item/3720-diez-lideres-de-antorcha-campesina-entre-los-propietarios-de-gasolineras-en-puebla-y-guerrero. Date accessed: July 13, 2021.

Dalby Chris. 2019. "Mexico's Santa Rosa de Lima Cartel Risks Burning Too Bright, Too Fast." *Insight Crime*, February 15. https://insigh tcrime.org/news/analysis/mexicos-santa-rosa-de-lima-cartel-el-marro/. Date accessed: July 16, 2021.

Dávila Patricia. 2019. "El Mapa del Huachicoleo." *Proceso*, January 27, 2019.

de Boer John and Bosetti Louise. "The Crime-Conflict Nexus: Assessing the Threat and Developing Solutions." Crime-Conflict Nexus Series: No 1. United Nations University Centre for Policy Research, pp. 1–12.

de Mauleón Héctor. 2015. "CJNG: La sombra que nadie vio." *Nexos*, June 1. https://www.nexos.com.mx/?p=25113. Date accessed: July 12, 2021.

de Mauleón Héctor. 2019. "El Cártel del Huachicol." *El Universal*, January 17. https://www.eluniversal.com.mx/columna/hector-de-mauleon/nacion/el-cartel-del-huachicol. Date accessed: July 12, 2021.

Edmundo Velázquez (Journalist and Director of Newspaper *Página Negra*) Interviewed by Samuel León in Mexico City, August 20, 2018.

Enciso Froylán. 2017. "México y la guerra sin nombre." *Crisis Group*, June 15. https://www.crisisgroup.org/es/latin-america-caribbean/mexico/mexicos-worsening-war-without-name. Date accessed: July 11, 2021.

EnergeA and Grupo Atalaya. 2017. "Estudio para analizar la problemática de seguridad física en las instalaciones del sector hidrocarburos y emitir recomendaciones para el reconocimiento de costos por concepto de seguridad que la Comisión Reguladora de Energía lleva a cabo en sus procesos de revisión tarifaria," pp. 1–368.

Farah Douglas. 2013. "Fixers, Super Fixers, and Shadow Facilitators: How Networks Connect." In *Convergence: Illicit Networks and National Security in the Age of Globalization*, edited by Michael Miklaucic and Jacqueline Brewer, 75–95. National Defense University Press.

Flores Pablo. 2019b. "Guanajuato alcanza en 2018 crecimiento de 4.4%." *Milenio*, January 4. https://www.milenio.com/politica/guanajuato-alcanza-2018-crecimiento-4-4. Date accessed: July 12, 2021.

France 24. 2017. "Oil Theft: A Multibillion-Dollar Business Fuels Mexican Cartels." Video. https://www.youtube.com/watch?v=ukJ7SWj4Wgk&t=1s. Date accessed: July 13, 2021.

Gabriel Stargardter (Journalist for Reuters) Interviewed by Samuel León in Mexico City, August 9, 2018a.

Garay-Salamanca Luis Jorge y Salcedo-Albarán Eduardo. 2016. "Macro-Criminalidad: Complejidad y Resiliencia de las Redes Criminales." iUniverse, pp. 1–191.

Garzón, Juan Carlos. 2014. "From Drug Cartels to Predatory Micro Networks: The 'New' Face of Organized Crime in Latin America." In *Reconceptualizing Security in the Western Hemisphere in the 21st Century*, edited by Bruce M. Bagley, Jonathan D. Rosen, and Hanna Kassab, pp. 117–131.

Gustavo Mohar (Risk Consultant and Former Intelligence Official) Interviewed by Samuel León in Mexico City, August 14, 2018.

Hübschle Annette. 2017. "Contested Illegality: Processing the Trade Prohibition of Rhino Horn." In *The Architecture of Illegal Markets: Towards an*

Economic Sociology of Illegality in the Economy, edited by Jens Beckert and Matías Dewey, 1–22. Oxford University Press.
Instituto Nacional de Geografía y Estadística (INEGI). 2018. "Datos Preliminares Revelan que en 2017 se Registraron 31 mil 174 Homicidios."
Jesperson Sasha. 2017. "Conflict Obscuring Criminality: The Crime-Conflict Nexus in Nigeria." Crime-Conflict Nexus Series: No 4. United Nations University, pp. 1–14.
Jiménez Benito. 2019a. "Toleraron el huachicol en Puebla." *Reforma*, February 18.
Jones, Nathan P. 2018. "The Strategic Implications of the Cártel de Jalisco Nueva Generación." *Journal of Strategic Security* 11 (1): 19–42.
Kary Siza y Los Chikos Kumbia. 2017. "La Cumbia Huachicol." Video. https://www.youtube.com/watch?v=gfYBWGMl7gA. Date accessed: July 13, 2021.
Llano Manuel and Flores C. 2017. "Ductos, ¿por dónde circulan los hidrocarburos en México?" [map]. Scale 1: 3,500,000. México: CartoCrítica/ Fundación Heinrich Böll.
Loret de Mola Carlos. 2019. "Luis Miranda en la Mira del Huachicol." *El Universal*, February 13. https://www.eluniversal.com.mx/columna/carlos-loret-de-mola/nacion/luis-miranda-en-la-mira-de-hacienda-por-huachicol. Date accessed: July 13, 2021.
Los Reyes Klan. 2017. "El Huachicol." Video. https://www.youtube.com/watch?v=pqFqcyYI3w8. Date accessed: July 13, 2021.
Martínez Mayra. 2015. "La crisis de Volkswagen pone en juego 78,000 empleos en Puebla." *El Economista América*, September 25. https://www.eleconomistaamerica.com/economia-eAm-mexico/noticias/7029460/09/15/Volkswagen-pone-en-juego-78000-empleos-en-Puebla.html. Date accessed: July 13, 2021.
Mastretta Sergio y Silva María Eugenia. 2019. "La Trama Audi Componendas de un Gobierno Autoritario." Puebla contra la corrupción y la impunidad, A.C., pp. 9–199.
Mastretta Sergio. 2017. "Escenas del Huachicol Poblano." *Nexos*, May 12. https://www.nexos.com.mx/?p=32348. Date accessed: July 13, 2021.
México Evalúa. 2018. "Homicidios y huachicol: un patrón en Guanajuato," October 26. https://www.mexicoevalua.org/homicidios-huachicol-patron-en-guanajuato-2/. Date accessed: July 12, 2021.
Noticieros en Línea. 2019. "Presunta Plática Entre Hugo Estefanía Y 'El Puma', Sobrino De 'El Marro'." Video. https://www.youtube.com/watch?v=2FhfzYQUMSY. Date accessed: July 13, 2021.
Noticieros Televisa. 2018. "¿Quién es el Bukanas?." Video. https://www.youtube.com/watch?v=gmHUgFreBS0. Date accessed: July 13, 2021.
Observatorio Nacional Ciudadano. 2018a. "Escenarios de Riesgo: Guanajuato." Observatorio Nacional Ciudadano, pp. 1–8.

Observatorio Nacional Ciudadano. 2018b. "Escenarios de Riesgo: Puebla." Observatorio Nacional Ciudadano, pp. 1–8.
Pérez Ana Lilia. 2018b. "Datos clave de la Refinería de Salamanca de Pemex." *Newsweek en Español*, December 28. https://newsweekespanol.com/2018/12/refineria-salamanca-huachicol-robo-gasolina-pemex/, Date accessed: July 12, 2021.
Reed Tristan. 2015. "Mexico's Drug War: A New Way to Think About Mexican Organized Crime." *Stratfor*.
Reina Elena. 2019. "José Antonio Yépez 'El Marro', el nuevo enemigo público de México." *El País*, March 10. https://elpais.com/internacional/2019/03/07/mexico/1551997956_195443.html. Date accessed: July 9, 2021.
Reyes Oscar. 2018. "Incrementan 225% homicidios en Corredor Industrial." *El Sol de Irapuato*, July 22. https://www.elsoldeirapuato.com.mx/local/incrementan-225-homicidios-en-corredor-industrial-1859221.html. Date accessed: July 12, 2021.
Robo de Combustible en Puebla 2000 a 2018. Gobierno Fácil. http://gobiernofacil.com/herramientas/robo-de-combustible-en-puebla. Date accessed: March 18, 2019.
Saucedo David. 2019. "Guerra de cárteles hunde al golpe de timón y baña de sangre a Guanajuato." *Poplab*, September 15. https://poplab.mx/article/GuerradecarteleshundealgolpedetimonybaadesangreaGuanajuato. Date accessed: July 9, 2021.
Secretariado Ejecutivo del Sistema Nacional de Seguridad Pública (SESNSP). "Datos Abiertos de Incidencia Delictiva." Open Data Base. https://www.gob.mx/sesnsp/acciones-y-programas/datos-abiertos-de-incidencia-delictiva?state=published. Date accessed: July 9, 2021.
Sergio Mastretta (Independent Journalist based in Puebla) interviewed by Samuel León in Mexico City, August 7, 2018.
Stargardter Gabriel. 2018b. "The Refinery Racket." *Reuters*, January 24. https://www.reuters.com/investigates/special-report/mexico-violence-oil/. Date accessed: July 12, 2021.
Stewart Scott. 2018. "Tracking Mexico's Cartels in 2018." *Stratfor*. February 1. https://worldview.stratfor.com/article/tracking-mexicos-cartels-2018. Date accessed: July 12, 2021.
Transparency Response "1857200119616." PEMEX. May 27, 2016.
Velázquez Edmundo. 2019a. "Los Bukanas, El Mamer, Sangre Nueva Zeta: estas son las bandas más peligrosas que operan en la autopista México-Puebla-Veracruz (MAPA)." *Página Negra*, January 28. https://www.periodicocentral.mx/2019/pagina-negra/delincuencia/item/1891-los-bukanas-el-mamer-sangre-nueva-zeta-estas-son-las-bandas-mas-peligrosas-que-operan-en-la-autopista-mexico-puebla-veracruz-mapa. Date accessed: July 13, 2021.

Velázquez Edmundo. 2019b. "Huauchinango, la nueva guarida de los huachicoleros." *Página Negra*, May 13. https://www.periodicocentral.mx/2019/columnistas/cuenta-hasta-10/item/10969-huachinango-la-nueva-guarida-de-los-huachicoleros. Date accessed: July 13, 2021.

Williams Phil. 2001. "Transnational Criminal Networks." In *The Future of Terror, Crime, and Militancy*, edited by John Arquilla, David Ronfeldt, 61–97. RAND Corporation.

Yashar Deborah J. 2018. *"Homicidal Ecologies: Illicit Economies and Complicit States in Latin America."* Cambridge University Press, pp. 1–368.

CHAPTER 6

The Mexican Fuel Black Market After a One Year Crackdown

INTRODUCTION

This chapter will assess how the MFBM reacted after the first year of a large-scale crackdown implemented by the Obrador government, which began in late 2018 and took place during 2019. It will be an opportunity to update Mexico's energy situation, as well as the policy implications of dealing with the MFBM. We will also delve into the current security situation in Mexico and develop the actions taken by the new presidency to tend to this pressing matter. This chapter will contribute to update the status of the MFBM during 2019, as well as to assessing the Mexican state's capacity to curb this criminal market. It will also allow to reflect on the resiliency of criminal networks, and to present some ideas encouraging further research down the road and speculate about the MFBM's future.

THE AMLO ENERGY CONTEXT

One priority has guided the Obrador government: the rescuing of PEMEX. Historian Lorenzo Meyer stated that Obrador's rescue project will determine the failure or success of his government, and the new economic model it intends to consolidate (Meyer 2019). The challenge of giving sustainability to this SOE is daunting. Declines from multiple fronts put into serious question the viability of PEMEX as an oil company.

© The Author(s), under exclusive license to Springer Nature Switzerland AG 2025
S. León Sáez, *Mexico's Fuel Trafficking Phenomenon*, St Antony's Series,
https://doi.org/10.1007/978-3-031-70503-8_6

Deteriorating reserves and dwindling production of crude and derivatives have become stablished trends of PEMEX's decline. For the last 15 years, overexploitation of reserves and the prioritization of extracting crude for export to meet PEMEX's fiscal obligations (historically representing 30–40% of government income) led to neglect in strategic areas like refining (Morales 2020, 18). In the meantime, the SOE's production structure followed an opposite path of Mexico's energy demand: while Mexico consumed more Liquified Petroleum Gas (LPG), diesel and gasoline, PEMEX's production focused on heavy crude and polluting fuel oil. This is serious considering these tendencies of internal energy consumption will continue for the foreseeable future.

Thus far, this production deficits have been covered by imports. Mexico is currently spending $19 billion USD a year on imported oil derivatives (Hernández and Bonilla 2020, 2). With the exchange fluctuations between the US Dollar and the Mexican Peso, these imports can have onerous costs for Mexico (Auditoría Superior de la Federación 2016, 5). In theory, it would be more cost-effective to produce derivatives than import them, though PEMEX's refining capacity made this impossible.

This decline in refining capacity finds its explanation in the fact that PEMEX prioritizes government agendas over technical objectives as an oil company, behaving as an income collector for the state. Even after the 2013 reform, the Mexican state remained dependent on income generated by this SOE. Recent declines in the amount of transfers to the government have had more to do with instability of global energy markets and PEMEX's decline than with measures redefining the company's role within the state. The Energy Secretariat still maintains a preponderant role in the SOE's board of directors and the Treasury still determines PEMEX's spending, budget and fiscal obligations.

With the rigidity brought upon its fiscal obligations, PEMEX has shown on repeated occasions that it lacks the flexibility to adjust its operational costs when production falls or the international market dwindles. Besides the challenges brought by declining production, growing imports and the burden of fiscal obligations, increasing exploration and production costs, inflated labour costs and massive debt (which led to PEMEX losing its investment grade in April 2020 (Nava 2020)), remain daunting challenges for the Obrador rescue plan. Because of these obstacles, in 2015 PEMEX's total expenses exceeded its total revenues. Profits in 2017 represented a meagre 7.5% of total sales income, while in the early 2000s they averaged 60% (Morales 2020, 12). In 2018 the SOE reported an annual loss of $7.6 billion (Alire 2019).

In early 2019 the PEMEX rescue plan was presented, resting on four pillars: (1) $15 billion MXN in annual tax breaks until 2024, (2) $25 billion MXN for capitalisation of the SOE, (3) $35 billion MXN from the payment of pension liability promissory notes and (4) $32 billion MXN ($1.6 billion USD) in savings from combating fuel trafficking. The presidency announced its crackdown on the MFBM as one of its main strategies to rescue PEMEX, promising billions of pesos in desperately needed resources. Additionally, this pledge was made in an urgently deteriorating security context.

The Obrador Security Context

The Obrador government in Mexico went into office facing a worsening security situation. In 2018 homicides broke a new record with 33,341 (Felbab-Brown 2019a, 6). To all the phenomena already explained, a more recent event took place aggravating criminal violence in Mexico: the expansion of the CJNG led to a conflict with the macro-criminal network of the Sinaloa Cartel. This conflict started during the second half of the EPN administration and that is still developing (Guerrero Eduardo 2018a, 2018b; Felbab-Brown 2019a, 5, 8). These two macro-criminal groups are disputing different black markets, in particular the control of trafficking chains of opiates and chemical precursors that feed demand in the US (Guerrero Eduardo 2018a, Felbab-Brown 2019a, 16). This event came to enhance criminal homicidal violence under the Obrador presidency.

As a response to this, the President opted for continuity regarding his security strategy, following the model of his predecessors of dismantling existing security institutions to create new ones relying on military personnel to carry out civil law enforcement (Felbab-Brown 2019b, 3). This is the case of the National Guard (NG), a militarized police force that is projected to reach 150,000 recruits. The NG is undergoing a process by which it will absorb the Federal Police (Guerrero Eduardo 2020), a security institution that could not be fully consolidated in 20 years (Hope 2020a).

The NG has been widely criticized for perpetuating the vertical, militarized security strategy followed by Calderón and EPN that has neglected the reform of local police forces (Felbab-Brown 2019b, 19), building civil policing models to develop intelligence-gathering capacities (Madrazo 2019), consolidating an effective justice system and associated to growing homicidal violence and human rights abuses in Mexico. A comparison between the 2018–2019 security budgets indicate continuity between the EPN and the AMLO administrations (Fig. 6.1).

Under Calderón and EPN substantial federal resources were transferred to municipal governments under the condition of professionalizing their police forces. According to Eduardo Guerrero (2020) "*this plan was a complete failure*" and, though some resources were spent on equipment, increased wages and control mechanisms for police forces, a vast amount was lavishly spent on public construction projects. Interestingly, this expenditure occurred in parallel to a militarization process of local police forces (Padilla Oñate 2019). As it is mentioned in the case study of Guanajuato, this militarization made local police forces more valued co-option assets, offering criminal groups access to heavily armed and well-equipped commandos (Saucedo 2019) (for some examples, see: Sources: SSPE; "Continúa reclutamiento nacional" 2019).

Ultimately, this poorly monitored spending "*failed to achieve transformative effects*" (Felbab-Brown 2019a, 13). Obrador's government responded to this issue reducing the Fund for the Professionalization

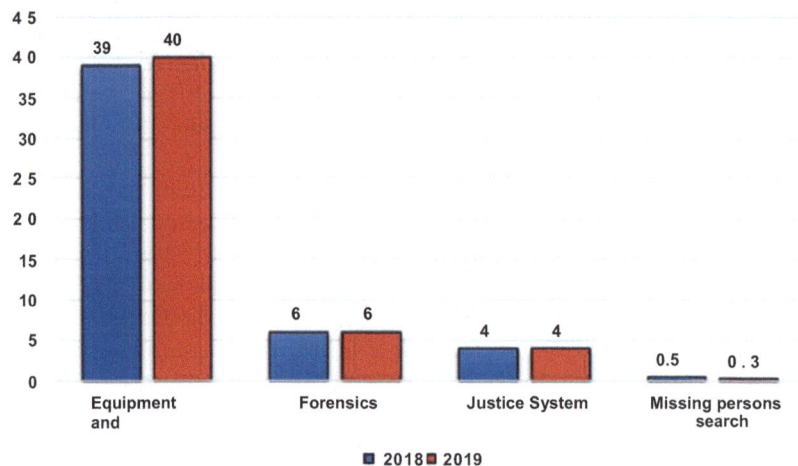

Fig. 6.1 Areas of security budget spending 2018 vs 2019 (%) ("La guerra que sigue" 2020)

and Equipping of the Municipal Police (FORTASEG) by 25% (Consensa 2019). While it is important to not misspend resources without clear goals or objectives, this measure makes the same omission as the president's predecessors. Just as throwing money at the problem did not work, this cut does nothing to alter the corrupt agreements between criminals and many state and municipal police forces in Mexico (Felbab-Brown 2019b, 19; Guerrero Eduardo 2020).

This is relevant considering the roles that police forces fulfil for criminal networks as grey actors at the local level. The participation of local law enforcers is a crucial element for local fuel-theft, as the state case studies of this research exemplify. This is also problematic because along *"with state police forces, the municipal police carry out most day-to-day policing"* (Felbab-Brown 2019b, 19). Therefore, local police forces are the first line of defence (or of collusion) against criminal cells and a potential crucial source to build a preventive intelligence-based policing model in Mexico.

A praise-worthy aspect is the commitment by the current administration to pursue illicit profits. In 2019, the amount of frozen bank accounts linked to illegal activities by the Mexican Financial Intelligence Unit (UIF) reached a record with over $5 billion MXN ($264.13 million USD) seized. This amount represented a 70-fold increase when compared to 2018 (*"La guerra que sigue"* 2020). But this positive policy pursued by the government falls short. First, it should be considered that *"chasing criminal money has not yet eliminated serious crime, such as violent drug trafficking, anywhere in the world"* (Felbab-Brown 2019b, 12) and that Colombia's criminality has thrived despite the country having some *"of the most stringent anti-money-laundering laws"* (ibid.).

Second, the centralized NG could prove of little use to seize ground-level criminal profits. This is because, as criminal markets become more fragmented, illicit actors have become increasingly local, operating largely with cash transactions. Without these groups being tackled by local law enforcement (engaging in community policing for intelligence gathering), they will find ways of sustaining their enterprises and revenues. Furthermore, of all the resources frozen by the UIF in 2019 a meagre 0.8% of the criminal cases presented before the Attorney General (FGR) were prosecuted (*"La guerra que sigue"* 2020). Obrador's security strategy, in its results and core policies, is not substantially different from those of his predecessors. But one detail distinguishes the president's security strategy: his "Hugs, not bullets" policy.

Of Hugs, State Lethal Force and Homicidal Violence

Unlike Calderón and EPN, AMLO has followed a distinct path regarding the use of lethal force by state forces. His two predecessors followed "*a logic of extermination*" (Guerrero Eduardo 2020) in which human rights abuses, lethal state force and criminal violence increased dramatically. Research has explained how the state can become a protagonist in escalating criminal violence. In a context of generalized repressive policies, criminal groups are incentivized to increase violence against state enforcers (Lessing 2015, 1499). If state crackdowns are widespread, violent conflict and lethality are likely to grow.

To this we can add the intra and interorganizational violence generated by the targeting criminal leaderships (Calderón et al. 2015, 5–6). Such strategies lead to increasing fragmentation, which expands the use of violence amongst expanding criminal competitors. Routine state violence destabilizes organizations leading to "*uncertainty about the strength of rivals and upstarts*" (Lessing 2015, 1506) promoting inter-criminal confrontation. Another consequence of all-encompassing coercion was the entrapping dynamic between criminal groups that pursued redirecting the repression of the state against their rivals (by blaming others for violence), a strategy known as "*calentar la plaza*".

For these reasons, experts on criminal violence advocate for more conditional approaches to state violence (Lessing 2015, 1507; Hope 2019; Kleiman 2011), a thesis ignored by Calderón and EPN with disastrous consequences for human rights. Human Rights Watch and Amnesty international denounced Calderón and EPN for implementing security approaches in which "*forced disappearances, extrajudicial executions, torture, and arbitrary detentions*" were commonly carried out by "*members of the military and by federal, state, and municipal police*" (Farfán 2019, 63–64). All this happened while both administrations made little progress in prosecuting abuses committed by security forces (Human Rights Watch 2017). During the past decade over 11,000 complaints for human rights abuses were presented before Mexico's Human Rights Commission (CNDH), while, according to data provided by SEDENA, between 2006 and October 2018 the military was involved in 4,272 confrontations with civilians (Rea and Ferri 2019, 22–23). This degrading human rights situation occurred in parallel to an increased use of lethal force by state actors.

Given the secrecy of Mexico's Defense Secretariat (SEDENA), the only data available on the use of lethal force by the military in public security matters covers the period between 2008 and 2014 (Farfán 2019, 64–65). During this time, between 10–15 civilians were killed for every fallen soldier. In 2011 that ratio of dead civilians per fallen soldier increased to an alarming 32.4 to 1. Overall, the lethality ratio of SEDENA increased from 1.6 to 1 in 2007 to 14.7 to 1 in 2012. The data of the Navy (SEMAR) also points to this trend of increasing state lethality: in 2004, 24 civilians were killed for every fallen marine; by 2014 that ratio had grown to 74 to 1 (ibid., 65). The Federal Police, on other hand, displayed ratios exceeding 10 dead civilians per fallen agent in 2011, 2012 and 2015 (Storr 2019, 101). Researchers who analysed this data concluded there was "*excessive and disproportionate use of lethal force as a possible pattern of behaviour of the federal forces*" (Silva et al. 2017, 355).

But after one year in office, AMLO seems to have broken with this concerning trend. The prominent phrase "Hugs, not bullets" has not only been a defining presidential discursive instrument, it has also shaped policy. The number of fallen civilians in confrontations with federal authorities had decreased by almost a third between January and September 2019 when compared to the same period during 2018 (Guerrero Eduardo 2020). It is the first time since 2010 that the number of detainees during armed confrontations with federal security forces exceeded those who were killed. If this is maintained, it will be one of the greatest accomplishments of this government; representing a turning point for the role played by the state in the Mexican security crisis as an important protagonist in reproducing lethal violence to an element of containment.

However, this decrease in the use of state lethal force has not been complemented by a reduction in criminal homicidal violence. The government expected that decreasing state lethality "*would suffice for criminal organizations to find ways to moderate their own violence*" (Guerrero Eduardo 2020). This was not the case.

Criminal Homicidal Violence Under AMLO

The night of August 28, 2019, a strip club was burnt in the port city of Coatzacoalcos, Veracruz. An armed commando arrived at the scene shooting anyone who crossed them. Afterwards, they covered the place in gasoline and sealed the exit once they set it on fire. Thirty people died that night, making it the deadliest crime-related massacre after 2011 when 52 people lost their lives in a fire provoked by the Zetas in a casino in Monterrey. This scene of horror is just one of 8 massacres that occurred during the first year of Obrador's government. In these attacks 124 people lost their lives, including children (Ángel 2019).

These massacres are symptomatic of a growth in the severity of criminal violence: homicidal incidents in 2019 took 1.17 lives on average, in 2018 and 2017 1.15, in 2016 1.11 and in 2015 1.10. In 19 out of 32 states homicides grew in 2019. Obrador's government argued that the rate of homicidal growth was not increasing as rapidly as it once was. During the second half of EPN's administration homicides grew at a rate of 17% while in 2019 that growth was of 2.7% (Hope 2020b). The problem with this argument is that this decrease was taking place since the last four months of the EPN presidency in 2018 and that by March 2020 the data pointed that the 2018 homicide monthly record (July with 3,158 victims) would be surpassed (ibid.). This is far from a turning point for homicidal violence in Mexico. Rather, the country is experiencing *"a stubborn stability in very high levels of homicidal violence"* (Hope 2020b).

What the failure of the hugs over bullets strategy seems to disturbingly demonstrate is that the criminal underworld has become so fragmented, competed and conflictive that the armed wings of criminal networks cannot restrict their violence. It also shows that the vicious cycle of increasing homicidal violence can maintain its high level even if the Mexican state assumes a containing rather than an enhancing role. 2019 closed with 35,551 homicide victims, a historical record that increased the homicide rate to 27 per 100,000 inhabitants (Beauregard 2020). In this extremely complicated public security context, the presidency would launch its crackdown and soon after would announce an astounding victory over the MFBM.

Triumphalism and the Crackdown(s) on the MFBM

In January 2019, the federal government launched a strategy to crackdown the MFBM under the name Joint Plan to Combat Fuel-theft. Mexico's military forces intervened in the country's 6 refineries, 39 storage and distribution terminals (including one marine terminal), 12 pumping stations and the SCADA pipeline monitoring unit in Mexico City. In total, 58 strategic facilities of PEMEX were intervened in 20 states across Mexico. Eight thousand six hundred security forces from

SEDENA, SEMAR and the NG were deployed to monitor PEMEX's pipeline system, with a presence of 50 security forces every 20 kilometres (Romero 2020).

According to PEMEX's Director, these actions translated into a 91% decrease of fuel-theft in polyducts from 20,400,000 barrels in 2018 to 1,800,000 barrels in 2019 (ibid.). This reduction represented savings of $56 billion MXN ($2.7 billion USD). This figure is 75% higher than the $32 billion MXN of savings linked to fuel-theft projected in the PEMEX rescue plan. For most of 2019 the Obrador Administration, facing meagre results regarding public security, was intensely promoting this decrease with a triumphalist discourse, stating pipeline fuel-theft decreased 95% (Rosas 2019). These numbers on a first instance seem impressive, yet further analysis is required to understand how the MFBM responded to the first year of a large-scale crackdown.

First, despite the scope of 2019's clampdown, this is not the first time the federal government has tried to tackle Mexican fuel trafficking. One of the most relevant examples of such interventions took place between March and October 2004, when the federal government implemented *Operativo Pemex* with the Federal Police spearheading the operation. More than 7,300 police agents were deployed across Mexico to takeover different energy installations and install highway checkpoints to find smuggled fuel. The installations, to which 1,300 agents were assigned to, included the six refineries, Storage and Distribution Terminals and polyducts in Estado de México, Guanajuato, Hidalgo, Jalisco, Nuevo León, Veracruz, Tamaulipas and Oaxaca (Medellín 2005; Vicenteño 2004a, 2004b; "Reditúa a Pemex" 2004). By August 2004, the government's intervention expanded to include 410 gas stations that were exhibiting discrepancies between the fuels they acquired from PEMEX and their sales (Vicenteño 2004b; "Rechazan gasolineros verificaciones" 2004).

The government at the time was not reluctant to victoriously publicize the results of *Operativo PEMEX*: $1.5 billion MXN ($94.5 million USD) were recovered, with sales increasing 11.5% after the first fifteen days of the federal forces' intervention ("Reditúa a Pemex" 2004). Yet, these successes were short-lived and by October 2005 intelligence reports of the federal government pointed to the existence of eight sophisticated criminal networks that trafficked oil derivatives in the northern, centre and western regions of Mexico (Medellín 2005). Another piece of data pointing to the dismal result of *Operativo PEMEX* is the variation of detected illegal extraction points in pipelines, which grew from 102 in 2004 to 132 in 2005 (Montero Vieira 2016, 3). What the MFBM's recent history points to is that anticipated victories should be taken with caution, and that crackdowns lead to temporary contractions followed by periods of adaptation and expansion (between 2004 and 2018 pipeline illegal extraction points grew uninterruptedly on a yearly basis). Adding to this, it is important to mention that *Operativo PEMEX* was not the only crackdown on the MFBM before the Obrador presidency.

In 2009 under Calderón, and during the fallout of the discovery of the Burgos Basin Operation, a clampdown of the federal government took place in which the Attorney General, the Public Security Secretariat, and the Federal Police took over PEMEX's Physical Security Services Management Unit ("Pemex investiga a..." 2009). The head of that unit (a Brigadier General) was detained for his involvement in fuel trafficking (ibid.). In 2011 the penalties for fuel-theft were increased, mandating prison sentences of 8 to 18 years (Pérez 2012). Despite this crackdown and increased punitive responses, the MFBM kept its growth during that period.

Under EPN, crackdowns on the MFBM also took place. In 2015, a security operation was launched by the Federal Police in which 700 agents backed by air support would patrol a polyduct between Ciudad Madero, Tamaulipas and the Cadereyta Refinery in Nuevo León ("Policía Federal despliega" 2015). In 2018 the Mexican Army and the National Gendarmerie (yet another national policing body formed under EPN) conducted a raid on Salamanca Refinery in Guanajuato. During this raid, law enforcement agents took documents and operational logs on the handling of fuels in that installation. Authorities looked for any information demonstrating how, between March and May of that year, gasoline and diesel theft increased through the alteration of measuring systems (Jiménez 2018).

What these examples show is that the MFBM is not unfamiliar to repression from the Mexican state, and that this black-market has found ways to maintain its continuity despite crackdowns. Without a doubt, the Obrador clampdown shines above the rest in its duration and resources. But still, these cases suggest cautious optimism, not anticipated triumphalism. This argument is further reinforced by the local dynamics displayed by the MFBM in 2019.

Of Local Dynamics, Inconsistencies and the Case Against Anticipated Triumphalism

Local dynamics are essential to interpret nation-level security trends. Illegal pipeline extraction points have indeed shown a decrease after the first year of the Obrador crackdown, but by not neglecting local dynamics we can see a more mixed scenario than simply an all-encompassing national-level drop (Table 6.1).

Table 6.1 Illegal extraction points in pipelines 2018–2019

State*	Jan–Dec 2018	Jan–Dec 2019	Variation (%)
Hidalgo	2,121	4,029	90%
Estado de México	1,517	1,778	17%
Puebla	2,072	1,846	−11%
Tamaulipas	1,301	1,153	−11.4%
Guanajuato	1,919	1,188	−38%
Veracruz	1,539	957	−38%
Jalisco	1,550	202	−87%
Querétaro	328	257	−21.6%
Tabasco	134	536	300%
Michoacán	205	284	38.5%
Tlaxcala	446	141	−68%
Oaxaca	183	140	−23.4%
Nuevo León	281	146	−48%
Baja California	142	164	15.4%
Sinaloa	425	77	−82%
Coahuila	33	69	109%
Chihuahua	132	60	−54.5%
Sonora	213	39	−81.6%
Aguascalientes	4	17	325%
CDMX	85	15	−82.3%
Chiapas	14	16	14.2%
Durango	36	13	−64%
Yucatán	4	6	50%
Morelos	209	4	−98%
San Luis Potosí	1	0	−100%
Total	14,894	13,137	

Elaborated with: Transparency Petition Responses 1857200007220 and 1857000059819
*States without Extraction Points in Pipelines 2018–2019: Baja California Sur, Campeche, Colima, Guerrero, Nayarit, Quintana Roo and Zacatecas

The data on the table above points to a national reduction of 11.8% in the number of illegal extraction points in pipelines between 2018 and 2019. It is far from a massive drop, but it is the first time since 2004 that this number decreases (Fig. 6.2).

This decrease shouldn't be overlooked nor exaggerated. If accurate, it is a decrease that represents a change in the growth of the MFBM, which had shown expansion for well over a decade. Some state decreases are notable. Yet what these pieces of data also show us is that the MFBM is dynamic and the growth in places like Hidalgo (where one of Mexico's six refineries is found), Estado de México and Tabasco (a key energy state) should remain points of concern. Also, the decreases in hotspots like Puebla and Tamaulipas, of around 11% each, indicate a long road ahead to significantly decrease the illegal commercialization of fuels, a challenge given the Mexican state's proven security limitations. Ultimately what the state variations in pipeline extraction show is that the MFBM combines cases of decreases and growth calling for cautious optimism, not the official triumphalism.

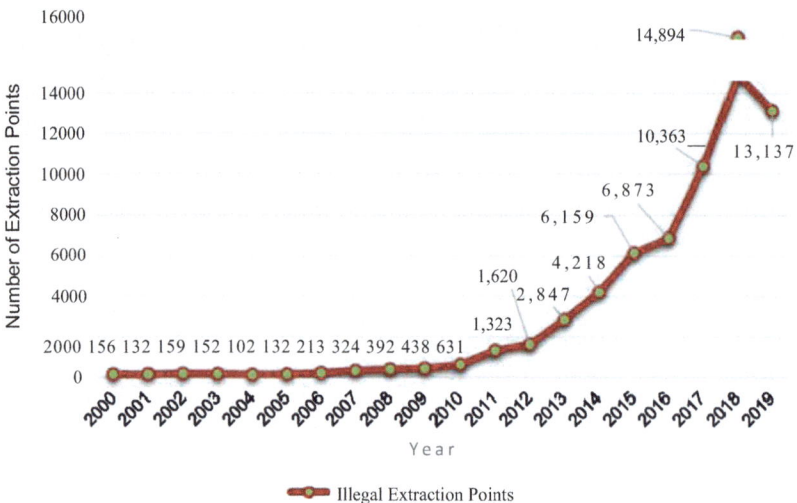

Fig. 6.2 Extraction points PEMEX Pipeline Network (PEMEX: "Reporte de tomas clandestinas"; EnergeA and Grupo Atalaya 2017, 344; Transparency Petition Responses 1857000059819 and 1857200094319)

Impunity is still an important incentive for the MFBM, with only 38.17% of illegal extraction points denounced before the Attorney General in 2019 (Table 6.2).

Table 6.2 Illegal extraction points in pipelines 2018–2019

State*	Jan–Dec 2019	Legal complaints presented before the Attorney General of the Republic (FGR)	Percentage (%)
Veracruz	957	894	93.4%
Hidalgo	4,029	727	18%
Puebla	1,846	652	35.3%
Estado de México	1,778	641	36%
Guanajuato	1,188	449	37.7%
Tamaulipas	1,153	344	29.8%
Nuevo León	146	317	217.1%
Tabasco	536	225	41.9%
Jalisco	202	127	62.8%
Oaxaca	140	116	82.8%
Michoacán	284	101	35.5%
Tlaxcala	141	89	63.1%
Querétaro	257	76	29.5%
Chihuahua	60	67	111.6%
Baja California	164	59	35.9%
Sinaloa	77	52	67.5%
Sonora	39	29	74.3%
CDMX	15	11	73.3%
Aguascalientes	17	10	58.8%
Coahuila	69	8	11.5%
Durango	13	6	46.1%
Morelos	4	6	150%
Yucatán	6	6	100%
Chiapas	16	2	12.5%
San Luis Potosí	0	1	–
Total	**13,137**	**5,015**	**38.17%**

Elaborated with: Transparency Petition Responses 1857200007220 and 1857000059819; IGAVIM (2020)
*States without illegal extraction points in 2019: Baja California Sur, Campeche, Colima, Guerrero, Nayarit, Quintana Roo and Zacatecas

As the data shows, the state with the largest number of illegal extraction points in pipelines in 2019, Hidalgo, also has one of the lowest percentages of legal complaints. This data not only shows that Mexico's pervasive impunity remains a key incentive for the functioning and resilience of the MFBM. It also shows the problems PEMEX faces with monitoring and measuring the exact number of illegal extraction points in its pipelines. As the table above shows, in Nuevo León, Chihuahua, Morelos and San Luis Potosí more legal cases are open concerning illegal extraction points than those that are detected.

The massive decreases of illegal extraction points experienced in Sinaloa and Jalisco should be subjected to further research, with the former state showing a drop of 82% and the latter a reduction of 87% between 2018 and 2019. For these cases it could be viable to hypothesize that the criminal macro-networks of Jalisco and Sinaloa retracted in their involvement in the MFBM in their main territories to avoid the presence of federal security forces in their strongholds. This doesn't mean these groups don't get involved (directly or indirectly) in other local-level MFBMs across the country using their loose network structures, it could be a matter of keeping their strongholds free from federal and military intervention. Related to this argument, on February 2019 CJNG signed messages appeared across the city of Guadalajara stating that this group would no longer be involved in fuel trafficking (Reza 2019).

Returning to 2019's illegal extraction points in pipelines, among state-level fuel trafficking hotspots the declines of Guanajuato and Veracruz are significant, both decreasing 38%. Yet in Guanajuato's case, the decrease of the MFBM has intensified black-market diversification's calamitous effect on homicidal violence. As the following graph shows, the decrease in the number of extraction points found in the state's pipelines between 2018 and 2019 transpired in parallel to an increase of almost 8% in criminal homicides (Fig. 6.3).

As it was mentioned in the case study of Guanajuato, the federal crackdown on fuel trafficking drove local criminals to diversify into predatory criminal enterprises (kidnapping, extortion and violent robberies) (Saucedo 2019). Think-tank *México Evalúa* criticised the federal government's fuel-theft strategy in Guanajuato for not effectively implementing contention strategies for the growth of homicidal violence in the state's municipalities where extraction points in pipelines decreased (López, Holst and Ramírez 2020). The reduction of pipeline theft combined with the fact that the state concentrates 20% of nationwide homicides (Silva

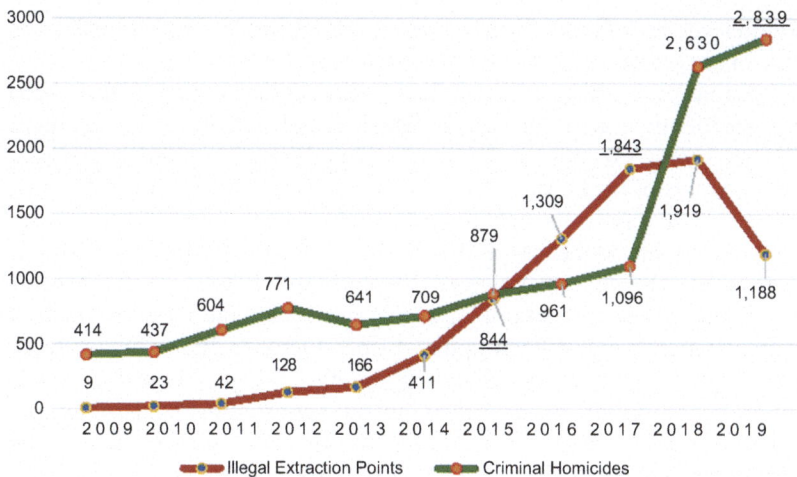

Fig. 6.3 Criminal homicides and illegal pipeline extraction points in Guanajuato 2009–2019 (*Source* SESNSP; López et al. 2020; Transparency Petition Responses 1857200007220, 1857000059819 and 1857200094319)

2020), serves as a cautionary example of the limitations of the Mexican state's security apparatus and it is another factor to consider when scrutinizing the celebratory posture of the presidency regarding the result of its crackdown on the MFBM.

One detail of the current governmental strategy presents itself as the most considerable shortcoming: reporting only drops in pipeline fuel-theft. As has been mentioned on this research, the MFBM is sustained by exploiting pipelines, refineries and storage and distribution terminals. Adding to this, theft can also take place through robbery of tanker trucks, fuel-transporting trains, piracy of tanker ships and theft of crude oil through bunkering between vessels and oil platforms (López Jorge 2019). This omission is even more troubling considering the importance that refineries and terminals have had for the MFBM according to the figures presented by PEMEX's Director in early 2019. The federal government is declaring victory on the MFBM based on incomplete data. Furthermore, there is the problem of the lack of control systems to monitor production levels in refineries and terminals (Romero 2019). Given all these arguments, we cannot overlook the work of journalists that shows, at the ground level, how fuel trafficking continued as a thriving and consolidated black-market in the middle of 2019 (Reforma 2019).

Sustaining the crackdown's success based on one source of fuel-theft, has coincidently gone unnoticed by Mexican public debate, though there have been some exceptions. *México Evalúa* criticized presenting this data as the sole proof of the success of the current crackdown on the MFBM, stating that it wasn't enough to make a clear assessment with decreases in pipeline theft alone (Moreno 2019). On top of this, this administration has shown reluctance to the publish of data on the MFBM on a regular basis. For example, a government online illegal extraction points database has not been updated since October 2018 ("Reporte de tomas" 2018). This leaves researchers, NGOs and analysts with the only option of getting information through time-consuming (and sometimes haphazardly answered) transparency petitions. This situation makes the independent analysis of the MFBM a difficult endeavour and does not allow for clear multi-sourced assessments to be devised.

Former energy official Rosanety Barrios argues that there are no elements demonstrating that the decrease in fuel-theft is having any impact on the SOE's finances (Meza 2020). During the first six months of 2019, when the President was already relentlessly promoting a massive decrease in pipeline fuel-theft, there was no effect of this decrease on the sales of PEMEX compared to the last six months of 2018, despite demand remaining virtually constant during that period (León 2019a). When reporting the financial results of these actions, the government showed significant inconsistencies. In January 2020, PEMEX declared that the savings emanated from their crackdown on the MFBM equalled $56 billion MXN ($2.96 billion USD). Yet this estimation was made considering the final sale prices to consumers (an average of $18.9 MXN/litre for gas station fuels during 2019). The problem here is that estimates in market value loss should be made with Pemex's first-hand sales prices which, when accounting for taxes and the margin for sellers, drop to $42 billion MXN ($2.23 billion USD). This $14 billion MXN ($741.1 million USD) difference becomes even more serious when considered that, during 2019, PEMEX sales losses dropped $99.2 billion MXN when compared to 2018 (García Karol 2020). This means that 2019's losses were 2.4 times higher than the supposed savings for the crackdown on fuel trafficking. Energy analyst Gonzalo Monroy criticized the Obrador administration for *"never materializing"* the expected results of their strategy against fuel-theft ("¿Cómo explicar la..." 2019). With all inconsistencies shown by authorities, this assertion seems accurate.

But these are not the only reasons to doubt the government's results of its clampdown on fuel trafficking. Data provided by PEMEX points to a considerably more complex situation regarding the MFBM after one year of the federal government's large-scale crackdown.

SITRAC and the Co-opted Institutional Reconfiguration of PEMEX

The Custody Transfer metering system (SITRAC) used by PEMEX measures fuel pipeline transportation from one operator to another. This system can help us clarify volumetric losses and it has been linked to fuel-theft in PEMEX on different occasions. SITRAC registries for example, were altered for theft back in 2005, modifying the number of hydrocarbons sent from the subsidiary *PEMEX Exploración y Producción* to the SOE's refining system (Pérez 2011). That same year, an audit by the governmental watchdog the *Auditoría Superior de la Federación* (ASF) used data of SITRAC to denounce that refining data provided by PEMEX had inconsistencies and was not reliable (ibid.). In 2007 Mexico's lower chamber of congress and the ASF detected, using SITRAC data, that PEMEX presented discrepancies equalling 31,387,000 barrels of oil between its figures regarding production, distribution and commercialization during 2005 (Garduño and Méndez 2007). A third example came in 2015, when PEMEX's 2014 Sustainability Report mentioned how the data provided by SITRAC allowed to locate illegal extraction points in pipelines across Mexico during that year ("Impera en 4..." 2015).

According to data from SITRAC, a conservative estimate points to 5,806,683 barrels of Magna and Premium gasolines, aviation fuel and diesel missing from different polyducts during 2019 (Transparency Response 1857000006220). This amount is considerably higher than the 1,800,000 barrels that were stolen in 2019 according to PEMEX's Director (Romero 2020) (Fig. 6.4).

From this total, 40% correspond to Magna gasoline, 28.8% to diesel, 27.2% to Premium gasoline and 3.8% to jet-fuel (Fig. 6.5).

These losses cannot be solely attributed to technical malfunctions. A revealing piece of data is that between 1994 and 2003 out of 441 incidents of fuel spillages in pipelines, 304 (69%) were attributed to fuel-theft. In a far second place, external corrosion caused 125 pipeline spillage incidents during that period (Leyva and Salazar 2017). What these numbers

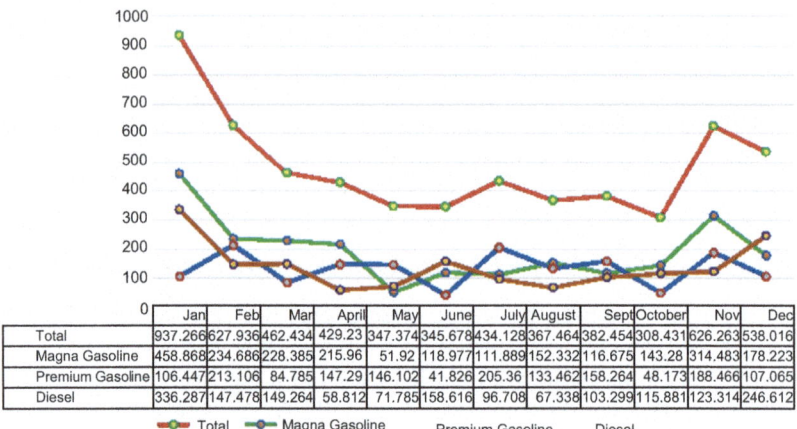

Fig. 6.4 Fuels missing on PEMEX's Custody Transfer metering system (SITRAC) 2019 (Thousands of Barrels) (*Source* Transparency Petition Response 1857000006220)

show is the importance fuel-theft has for pipeline recorded losses and the lower impact of technical malfunctions. Noé Cruz, an energy sector journalist with 30 years' experience, denounced that massive losses recorded by SITRAC point to "*wholesale fuel-theft operations*" (Cruz Serrano 2020) targeting PEMEX and that this necessarily involves the complicity of company insiders "*who know the type of product being transported, the time and the pipeline*" (ibid.).

The volumes of lost barrels registered by SITRAC are staggering. Cases where tens of thousands of barrels vanish from pipelines are common and, in some cases, these losses can reach hundreds of thousands of barrels, including massive fuel deliveries that disappear completely. The following are selected cases of losses recorded by PEMEX's Custody Transfer metering system reflecting these scenarios (Tables 6.3 and 6.4).

If we consider that during 2019 the average prices for Magna gasoline, Premium gasoline and diesel were $19.35 MXN/litre ($1.02 USD), $20.78 MXN/litre ($1.1 USD) and $21.16 MXN/litre ($1.12 USD) respectively, and that each barrel contains 159 litres of fuel, we can grasp how significant these losses are. What these figures also reflect is one of the largest omissions of the current crackdown on the MFBM: the federal government has not articulated a large-scale strategy to tackle

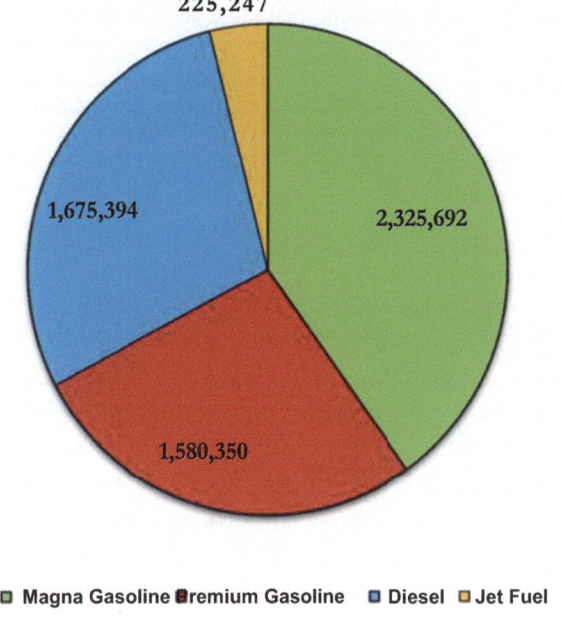

Fig. 6.5 Fuel types missing on SITRAC system 2019 (Barrels) (*Source* Transparency Petition Response 1857000006220)

the phenomenon of *Systemic Macro-Corruption* (Garay-Salamanca et al. 2018, 30–31) within PEMEX and key grey actors outside the SOE. As it was mentioned before in this research, the diversity of grey actors involved in the MFBM and the complexity of their operations drove to a Co-opted Institutional Reconfiguration (CItR) of PEMEX. Although the Obrador administration has shown willingness to tackle co-optation among some high-level officials (Rodríguez 2019), these actions appear to be isolated cases and there is no institutional-scale strategy to tackle this phenomenon of Macro-corruption which has permeated all of PEMEX's hierarchy (León 2019b).

After PEMEX's internal fuel sales declined 17.5% in 2019 and almost doubled its overall losses ("Pérdidas de Pemex" 2020), Ana Lilia Pérez declared that within PEMEX many groups still involved in fuel trafficking were operating despite federal government efforts. Pérez added that the SOE lacked controls on production measurements and that this

Table 6.3 Selected cases of missing fuel SITRAC system 2019

Month	Polyduct	Product	Quantity sent (Thousands of Barrels)	Quantity received (Thousands of Barrels)	Difference (Thousands of Barrels)
January	Polyduct 14" Satélite-Gómez Palacio	Diesel	150.071	5.013	−145.058
January	Polyduct 16-14-14" Tuxpan-Poza Rica-Tula	Premium Gasoline	159.262	89.523	−69.739
January	Bidirectional Polyduct 12-10" Madero-Cadereyta	Magna Gasoline	146.594	91.806	−54.788
February	Polyduct 16" Tula-Azcapotzalco	Magna Gasoline	26.568	0.000	−26.568
February	Polyduct 24-18-14" Tuxpan-Poza Rica-Azcapotzalco	Premium Gasoline	627.672	555.470	−72.202
February	Bidirectional Polyduct 12" Cd. Juarez-Chihuahua	Diesel	118.393	47.557	−70.836
March	Bidirectional Polyduct 12-10" Madero-Cadereyta	Magna Gasoline	273.205	194.384	−78.821
March	Polyduct 18" Cadereyta-Satelite	Diesel	730.555	675.894	−54.661
March	Polyduct 16-14-14" Tuxpan-Poza Rica-Tula	Premium Gasoline	467.634	448.627	−19.007
April	Bidirectional Polyduct 12-10" Madero-Cadereyta	Premium Gasoline	165.879	98.745	−67.134
April	Polyduct 14" Satélite-Gómez Palacio	Magna Gasoline	170.611	99.736	−70.875
April	Polyduct 10" Rosarito-Ensenada	Diesel	80.118	66.396	−13.722
May	Polyduct 14" Satélite-Gómez Palacio	Premium Gasoline	30.105	0.000	−30.105
May	Polyduct 10" Salamanca-Morelia	Diesel	113.602	95.433	−18.169

(continued)

Table 6.3 (continued)

May	Poliducto 14" Satélite-Gómez Palacio	Magna Gasoline	461.051	440.764	−20.287
June	Poliducto 16" Salamanca-Guadalajara	Diesel	321.166	266.742	−54.424
June	Poliducto Bidireccional 12-10" Madero-Cadereyta	Magna Gasoline	243.949	197.143	−46.806

reflected a long running technological lag within the company. Finally, she mentioned that the number of criminal elements involved, which include "*agents and former agents of the state, private companies, as well as the use of front companies and forged invoices*" ("En Pemex se" 2020) required a large-scale criminal trial similar to that of the *Maxiprocesso* in which 475 *mafiosi* were indicted in Italy in 1986. What all this information points to is that the CItR and Systemic Macro-corruption are still present in PEMEX.

On the issue of PEMEX's underdeveloped technologies, we must also reconsider the limitations of the monitoring system SCADA, which can only close 38% of the valves of the entire system remotely, and those of the monitoring sites, with 45% not working due to lack of repair, poor installation or not being installed at all (Flores 2019). Guesstimating has been a denominator for Obrador's administration since the first steps were taken to tackle the MFBM and evidence shows that the Mexican government lacks the capacity to measure this black-market precisely. This is a problem derived from a larger one: the Mexican state not only lacks a monitoring and measuring capability in regard to the MFBM, it doesn't have the capacity to tackle it efficiently. And this takes us to the closure of pipelines.

Closing Pipelines: A Costly Solution

An integral part of the government's crackdown has been to close pipelines to stop theft. In December 2018, the Obrador presidency closed polyducts all across the nation to stop pipeline fuel-theft, amongst them were the Tuxpan–Tula polyduct, the Brownsville–Reynosa–Cadereyta, the Madero–Victoria–Cadereyta, the San Martín Texmelucan–Valle de

Table 6.4 Selected cases of missing fuel SITRAC system 2019

Month	Polyduct	Product	Quantity sent (Thousands of Barrels)	Quantity received (Thousands of Barrels)	Difference (Thousands of Barrels)
June	Poliducto 16-14-14" Tuxpan-Poza Rica-Tula	Premium Gasoline	556.802	544.199	−12.603
July	Poliducto 10" Satelite - Monclova - Sabinas (Ducto Alterno)	Magna Gasoline	29.824	0.000	−29.824
July	Poliducto 16-14-14" Tuxpan-Poza Rica-Tula	Premium Gasoline	460.578	343.581	−116.997
July	Poliducto 18" Cadereyta-Satelite	Diesel	666.542	622.552	−43.990
August	Poliducto 14" Satélite-Gómez Palacio	Diesel	89.310	40.594	−48.716
August	Poliducto 24-18-14" Tuxpan-Poza Rica-Azcapotzalco	Premium Gasoline	770.745	732.371	−38.374
August	Poliducto 10" Satelite - Monclova - Sabinas (Ducto Alterno)	Magna Gasoline	105.152	68.028	−37.124
September	Poliducto 14" Satélite-Gómez Palacio	Magna Gasoline	310.444	259.775	−50.669
September	Poliducto 16-14-14" Tuxpan-Poza Rica-Tula	Premium Gasoline	502.680	380.827	−121.853
September	Poliducto 24-18-14" Tuxpan-Poza Rica-Azcapotzalco	Diesel	130.168	79.361	−50.807
October	Poliducto 18" Cadereyta-Satelite	Diesel	741.197	703.139	−38.058
October	Poliducto 24-18-14" Tuxpan-Poza Rica-Azcapotzalco	Magna Gasoline	3,163.415	3,083.872	−79.543
October	Poliducto 16" Salamanca-Guadalajara	Premium Gasoline	131.619	113.366	−18.253
November	Poliducto 16" Tula-Salamanca	Premium Gasoline	327.451	256.265	−71.186
November	Poliducto Bidireccional 12" Cd. Juarez-Chihuahua	Magna Gasoline	212.111	151.883	−60.228

(continued)

Table 6.4 (continued)

Month	Polyduct	Product	Quantity sent (Thousands of Barrels)	Quantity received (Thousands of Barrels)	Difference (Thousands of Barrels)
November	Poliducto 24-18-14" Tuxpan-Poza Rica-Azcapotzalco	Diesel	268.165	219.926	−48.239
December	Poliducto 16-14-14" Tuxpan-Poza Rica-Tula	Magna Gasoline	1,047.333	968.454	−78.879
December	Poliducto 18" Cadereyta-Satelite	Diesel	542.817	466.361	−76.456
December	Poliducto 16-14-14" Tuxpan-Poza Rica-Tula	Premium Gasoline	457.732	396.243	−61.489

Source Transparency Petition Response 1857000006220

México, Turbosina Tula–Azcapotzalco, Tula–Salamanca and the Salamanca–Guadalajara (Carretto 2019). By mid-January 2019, large-scale shortages were happening in Aguascalientes, Coahuila, Chihuahua, Durango, Guanajuato, Guerrero, Hidalgo, Jalisco, Ciudad de México, Estado de México, Michoacán, Nuevo León, Oaxaca, Puebla, Querétaro and Tamaulipas. In the states of Jalisco and Guanajuato shortages were such that only 15% and 10% of all state gas stations were operating respectively. The closure of pipelines was such that tankers were anchored in the Mexican Gulf coast because their fuel shipments could not be transported inland once they reached PEMEX's marine terminals ("Varados, 24 buques" 2019).

The economic impacts of these shortages were far-reaching. According to Credit Suisse, the losses associated to pipeline-closure were estimated at $5.6 billion USD (de la Rosa 2019). A study by the Jesuit University of Guadalajara estimated this loss at over $60 billion MXN ($3.13 billion USD) during January–February 2019 (Valdivia 2019). These costs outweighed in a couple of months the benefits of decreasing the MFBM that, in its 2018 peak, represented losses for PEMEX between $2.5–$3.5 billion USD (Raziel 2018). Though supply was subsequently normalized, closure of some pipelines remained into 2020.

By January 2020 PEMEX made public their plan to reopen 5 polyducts during that year: the Tula–Toluca, San Martín–Puebla, Minatitlán–Salina Cruz, Matamoros–Cadereyta and the Minatitlán–México. 220 kilometres of the Tula–Toluca polyduct have been closed since February 2019 because of the large amount of illegal extraction points, 225 kilometres of the Matamoros–Cadereyta polyduct remained closed though it is unknown since when (Hernández 2020; Romero 2020). Amongst the currently operating polyducts, 306 kilometres of the Tuxpan–Tula has been identified as the most conflicting section of that pipeline because of the number of illegal extraction points discovered in it. Energy specialists have stated that this polyduct has not been closed because doing so would generate shortages in Mexico's centre region (ibid.).

To make up for the closed pipelines, the federal government created a land transportation group controlled by the armed forces. This group is integrated by 612 tanker trucks, 1,681 drivers and armed convoys. Through this group PEMEX affirmed transporting 1.7 billion litres of fuels during 2019 (Romero 2020). The Mexican Association of Gasoline Entrepreneurs (AMEGAS) expressed concern over the closure of pipelines considering that tanker truck transportation is 14 times more expensive. AMEGAS also criticized the viability of sustaining fuel transportation through tanker trucks considering that the capital of Jalisco, Guadalajara, would require 400 trucks alone to feed its energy demand (González 2019). The total of tanker trucks in the hands of private actors and *PEMEX Logística* total 1,485 which have a carrying capacity of only 12% of Mexico's daily fuel consumption (ibid.).

We must turn our attention again to the local level and ponder the consequences of reopening certain pipelines without contention measures to guarantee the security of those areas and their populations; a cautionary example of this is the case of the escalating homicidal violence in Guanajuato. But if Guanajuato's case points to the risk of reducing the profits of the MFBM, in other places the concern is about the return of the fuel trafficking criminal cells as soon as the pipelines were reopened. In San Martín Texmelucan, Puebla, a municipality that concentrated almost 40% of illegal extraction points in that state in 2018 (Gobierno Fácil Puebla 2020), news outlets pointed to the concern of local communities over the reopening of the polyduct that crosses that area and its impact on increasing homicidal violence ("Reapertura de poliducto" 2020). Local authorities declared that they had not been contacted by PEMEX as to how the strategy would guarantee security once these installations were reopened (ibid.).

Because of security and technical limitations, the state crackdown on the MFBM opted for costly measures at the expense of energy strategic installations and the security of local populations. After a year of these efforts, evidence indicates that this black-market is thriving in local hotspots while decreases in others have remained stubbornly low. And still, its results lack the proper basis to make a comprehensive assertion of the true outcomes of this clampdown. In this regard, PEMEX should publish statistics in a timely manner, elaborate a progress report with sanctions, operation costs, recovered fuels, their impact on the SOE's finances, their usage and the progress made in its cooperation with other institutions (SEDENA, SEMAR and the National Guard) and so on. The situation, considering all the factors mentioned above, calls for cautious optimism at best, not anticipated triumphalism. And this whole situation becomes even more complicated once we consider the resiliency of criminal networks.

The Resiliency of Mexican Criminal Networks

The state has considerable limitations to tackle the MFBM and other criminal enterprises. Moreover, effective state interventions have remained elusive because of the resilience that Mexican criminal networks have shown since the beginning of the country's public security crisis. Since then, criminal networks have endured and adapted to changing conditions due to competition, changes in policy and law enforcement priorities, technological disruption and the expansion or contraction of illicit markets (Ayling 2009, 182).

In this regard, adaptability, understood as the capacity to turn "*challenges into opportunities*" (ibid., 184), is a key element for the resiliency of criminal networks. The migration towards new territories to continue trafficking hydrocarbons (as the state shifts in pipeline extraction show) is one example of this resilience and adaptability. It also attests to the argument that criminal networks cells operate without large infrastructure investments, making them mobile.

The migration of the operations linked to the MFBM can also be witnessed within states. Local journalists have mentioned how in Puebla, fuel trafficking cells quickly changed their operations to exploit polyducts in isolated areas in the Sierra Norte de Puebla, taking advantage of the difficult terrain, the impoverished communities and absence of authorities (Velázquez 2019). Infrastructural state capacity can vary considerably within national territories. This includes the state's capability to maintain a presence and deploy security forces across its geography, a matter that has proved a great challenge for the Mexican state during the security crisis. Within territorial enclaves, criminal groups will tend to operate with greater capacity in areas where the presence of security forces is functionally inactive and/or vulnerable to co-option. These groups will evade the state and exploit its uneven territorial display capability. Furthermore, the efficiency in which criminal cells migrate their operations shows that the criminal networks involved in the MFBM can quickly adapt to the disruption of the government's crackdown (Ayling 2009, 185).

Co-option of legal business and state authorities provide sources of resilience to disruptions (Ayling 2009, 189). Legal businesses can protect illicit resources from state crackdowns and authorities can protect corrupt agreements sustaining impunity. Redundancy is another resilience source, a network structure with many redundant nodes "*will have less difficulty adapting to the removal of actors*" (ibid., 190). Keeping peripheral nodes redundant insulates their core counterparts. This allows to protect the information and resource-concentrating elements of the network that can reorganize after disruption.

Mobility, adaptability, co-option and redundancy are key elements of criminal resiliency, but they are not the only ones. The various illicit opportunities of the Mexican context enhance the resiliency of its criminal networks. If a black-market opportunity closes, new options for profits are constantly available, increasing diversification and options for adaptation. And, as we will see in the next section, the MFBM has not neglected these options while confronting the Obrador crackdown.

The Diversification of the Fuel Black-Market

Diversification towards new illegal enterprises can help criminal networks to adapt to external pressures (Fuerte Celis et al. 2018, 5). Consequently, diversification can be understood as a means for adaptation. With the pressures emanating from the federal government's clampdown on fuel trafficking, criminal groups diversified into another branch: Liquified Petroleum Gas (LPG). According to private companies of this sector, the theft and illegal commercialization goes back to the year 2010 (Cacho 2019, 38). However, as the presidency began pursuing the theft of fuels like diesel and gasoline, criminal networks branched out into this widely used hydrocarbon. By late 2019 the Mexican Association of Liquefied Gas Distributors (AMEXGAS) stated that criminal networks involved in LPG trafficking represented an illegal market of $13 billion MXN ($670 million USD) (Nájar 2019). Besides targeting PEMEX, this black-market also exploits private distributors.

The CJNG, the Sinaloa Cartel, the SRLC, the Zetas and the Cártel del Noreste (a Zetas split-group present in Tamaulipas, Nuevo León and Coahuila) are involved in exploiting this new branch of the MFBM concentrated in industrialized hubs of the country ("Los Sinaloa y" 2019; "El nuevo reto" 2019; Sierra 2019). With the involvement of these and other smaller-sized criminal networks, the theft and trafficking of LPG has expanded: between 2018 and 2019, the number of illegal extraction points for LPG pipelines increased almost 497% (Table 6.5).

Puebla concentrated over 69% of all LPG extraction points, outperforming other gas-theft hotspots significantly. Within the state, the municipalities with the highest number of illegal LPG extraction points also had a history of being fuel trafficking hotspots: Tepeaca, Texmelucan, Los Reyes de Juárez, Amozoc, Palmar de Bravo and Acajete (Jiménez 2019a). Other state MFBM hotspots have demonstrated diversification towards the criminal exploitation of LPG during 2019 (Guanajuato, Estado de México, Hidalgo, Veracruz). Puebla and Estado de México concentrate 88.6% of all LPG extraction points detected during 2019.

Table 6.5 Illegal extraction points in LPG pipelines 2018–2019

State	2018	2019	Variation (%)
Puebla	64	918	1,334.37%
Estado de México	93	256	175.26%
Tlaxcala	9	40	344.44%
Veracruz	5	28	459.99%
Querétaro	20	26	30%
Hidalgo	6	26	333.33%
Nuevo León	16	18	12.5%
Guanajuato	9	10	11.11%
Jalisco	0	2	–
Tabasco	0	1	–
Total	222	1,325	496.85%

Source IGAVIM (2020)

Sector representatives affirmed that 15 to 20% of monthly LPG consumption in Mexico City, Estado de México, Puebla, Veracruz, Hidalgo, Tlaxcala and Guanajuato was under black-market control in 2019 (García Karol 2019). It is estimated that this number stands for 8% of all the LPG commercialized in Mexico monthly (60,000 of a total of 750,000 tons) (Nava 2019).

Criminal diversification has taken place in response to the crackdown of the federal government that prioritized targeting the theft of fuels such as gasolines and diesel ("Puebla, una de" 2019). This is an act of resilience, understood as the ability of criminal groups *"to absorb disturbance and reorganize while undergoing change so as to retain the same function [and] structure"* (Ailyng 2009, 183). The criminal networks of the MFBM have responded to the clampdown finding opportunities in adversity. And as we will see, there are sound reasons to diversify to LPG.

The Motivations to Diversify Towards LPG

Criminal diversification towards LPG has not been fortuitous. Criminal networks have rationally boosted their involvement profiting out of this fuel because of several advantages. First, the gas pipeline network (transporting natural gas and LPG) has a considerably larger span of 34,242.97 km compared to polyducts, which covers 9,098.53 km (Llano and Flores 2017) (Map 6.1).

Pipelines cover a larger territory and have more actors involved in them than the PEMEX-managed polyducts. Private companies, the Federal Electricity Commission (CFE) and PEMEX participate in this network, making it more complex and diverse (Leyva and Salazar 2017, 41). The larger extension and the fact that private actors operate their own pipelines has made this a more appealing alternative than profiting from fuel trafficking derivatives like gasoline and diesel (which has been the priority of the government crackdown). There are five gas pipelines operating in Mexico: the PEMEX owned SNGLP Cactus-Guadalajara and Hobbs-Méndez and the privately managed Penn Octane, the Reynosa Monterrey pipeline and *Ductos del Altiplano*. Out of these, the SNGLP is the largest with a carrying capacity of 67% of the entire network (Madrid Ayala et al. 2018, 40).

The SNGLP pipeline, passing through Tabasco, Veracruz, Puebla, Hidalgo, Estado de México, Querétaro and Jalisco, has the largest number of illegal extraction points according to PEMEX (Jiménez

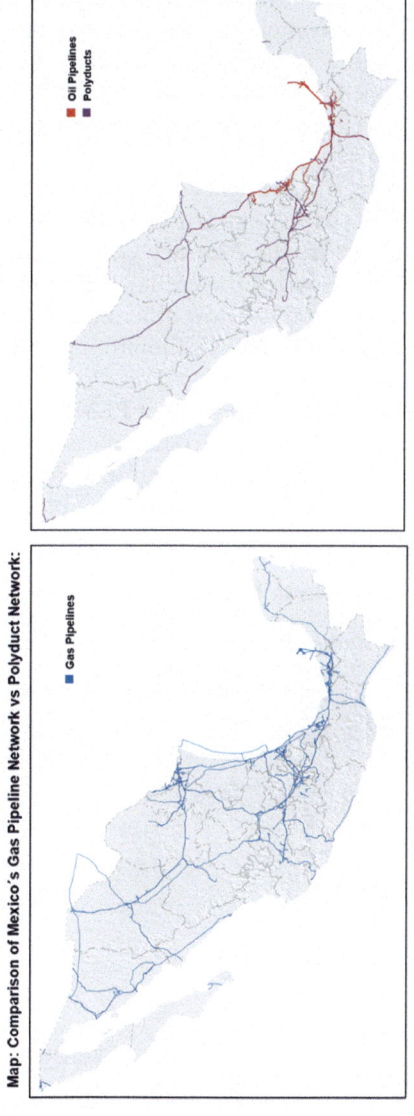

Map 6.1 Comparison of Mexico's gas pipeline network *vs* Polyduct network (Llano and Flores 2017)

2019a). AMEXGAS estimates that $5.58 billion MXN ($294.5 million USD) worth of LPG was subtracted from the SNGLP pipeline during 2019 (García Karol 2019). The extension and the diversity of involved actors in this network further complicates the monitoring and protecting from the state. Adding more complication, the flow of LPG in these pipelines cannot be stopped because certain industries depend on its constant flow (ibid.). Therefore, it is not possible to replicate the costly polyduct closure strategy.

Lacking law enforcement is a deficiency incentivizing LPG trafficking. In 2019, only 17.13% of illegal extraction points in PEMEX LPG pipelines were denounced before the Attorney General (Table 6.6).

This situation further incentivizes the criminal diversification towards LPG: prosecution levels considerably below other branches of the MFBM. And yet again inconsistencies in PEMEX's monitoring and the presented legal complaints arise, as the cases of San Luis Potosí and Michoacán show, where complaints were presented without PEMEX registering them as extraction points. Veracruz is another troubling case, where the detection of illegal extraction points does not translate into a single legal complaint. Additionally, LPG is not stolen only from pipelines. The theft

Table 6.6 Illegal extraction points in PEMEX LPG pipelines 2018–2019

State	Jan–Dec 2019	Legal complaints presented before the Attorney General of the Republic (FGR)	Percentage (%)
Puebla	918	153	16.67%
Estado de México	256	28	10.94%
Nuevo León	18	14	77.78%
Tlaxcala	40	12	30%
Guanajuato	10	9	90%
Querétaro	26	4	15.38%
Hidalgo	26	5	19.23%
San Luis Potosí	0	1	–
Michoacán	0	1	–
Veracruz	28	0	0%
Jalisco	2	0	0%
Tabasco	1	0	0%
Total	1,325	227	17.13%

Source IGAVIM (2020)

Table 6.7 Legal complaints presented before the Attorney General of the Republic (FGR) for LPG tanker truck theft

State	Complaints
Puebla	126
Querétaro	7
Hidalgo	7
Baja California	3
Tlaxcala	2
San Luis Potosí	1
Nuevo León	1
Estado de México	1
Total	148

Source IGAVIM (2020)

of tanker trucks is another means for criminal networks to access this fuel (Table 6.7).

It is estimated that 38.1% of all vehicle theft in Mexico is not reported to authorities (INEGI 2019) and that representatives from private companies declared that they had over 400 tanker trucks stolen during 2019 ("Puebla, una de" 2019), so these figures are undoubtedly higher. Tanker trucks are not stolen to access the LPG once, they are also used for continued trafficking and commercialization. Journalistic sources in Puebla have documented the existence of criminal cells specialized in stealing tanker trucks to sell to LPG traffickers for $25,000–$30,000 MXN each ($1,287–$1,544.5 USD) (Zenteno 2020).

As with other hydrocarbons, black-market LPG operations rely on illegality camouflaging. In Hidalgo, Puebla and Mexico City criminal networks have been found camouflaging delivery vehicles (in individual tanks) and tanker trucks to make them appear like they belong to legitimate companies (Nájar 2019). In the municipality of Texcoco in Estado de México the company *Gas LP México* has been investigated by authorities for functioning as a front to distribute stolen LPG (Jiménez 2019b). LPG distribution in Mexico "*is an activity in which Pemex does not participate and is carried out exclusively by private companies*" (Madrid Ayala et al. 2018, 49). The participation of private actors in LPG distribution makes it more difficult to identify illicit agents in its commercialization.

In the case of Texcoco, grey actors have also been implicated. Companies who tried to commercialize LPG in Texcoco, and had their drivers attacked by armed groups, declared that these cells operate with the support of a former major of that municipality who is now a federal official and that the current municipal government is also involved with LPG traffickers. Local businessowners complained about the inaction of the local police to tackle these armed cells (Jiménez 2019b). LPG trafficking operates relying on illegality camouflaging and on grey actors, just as it occurs with the other MFBM operations (Illustration 6.1).

76% of Mexican households use LPG for cooking and 35.5% for water heating (Madrid Ayala et al. 2018, 26). Most of these households belong to lower income segments, which have higher expenditure levels for this fuel (ibid., 27). And this is yet another incentive of the LPG black market: it allows criminal networks to profit from lower-income segments of Mexican society. In lower income areas state presence is weak, corruption and illicit opportunities are more prevalent and there are widespread informal economies facilitating money-laundering (Durán-Martínez 2018, 6–7; Radden Keefe 2013, 106).

The LPG sector in Mexico has had historical informal branch. This is true not only for Mexico but also for Latin America, where LPG informal markets are prevalent everywhere except in Chile (López 2012). In 2012

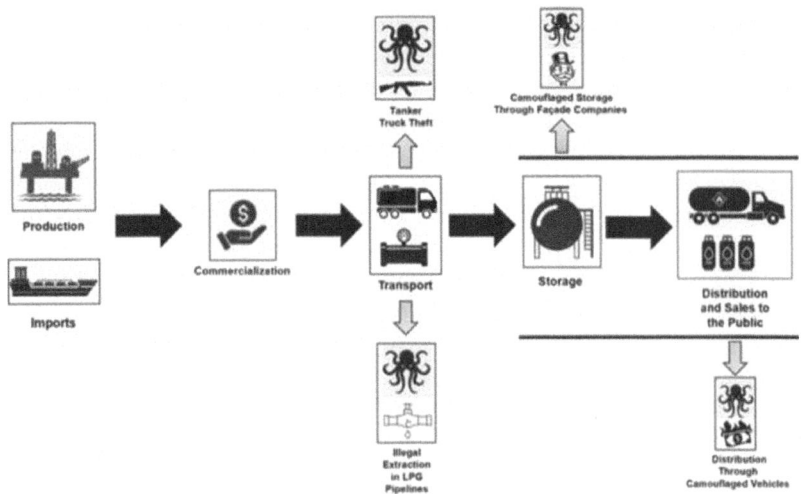

Illustration 6.1 Liquified petroleum gas value chain and points of criminal network intervention (Elaborated by the author with information from Madrid Ayala et al. 2018, 30)

the Iberoamerican Association of LPG (AIGLP) warned that Mexico had the largest informal LPG sector in Latin America and that it was "*the backbone of the stolen LP gas*" market (ibid.). This informal market consolidated because of historical price controls that lasted until 2017 (Madrid Ayala et al. 2018, 74). The rigidity of these controls hindered LPG supply in areas where their populations lacked the purchasing power to consume this fuel. This void created by price controls was filled by informal distributors (CIDAC 2015, 21).

Price liberalization expanded the formal market, yet it did not make its informal counterpart disappear. In 2018 the Federal Commission on Economic Competition (COFECE) alerted about economic agents that operated between LPG distributors and final consumers. These agents, known as commission agents, were "*informal independent intermediaries*" (Madrid Ayala et al. 2018, 62) who can play roles as either clients of distributors or as their informal competitors. COFECE warned that the presence of informal (and sometimes illegal) agents in the distribution of LPG could hinder the entrance of new competitors into this market.

These circumstances facilitated the diversification of illicit networks to develop an LPG criminal market. Criminal networks, for example, found informal distributors that were easy to co-opt and a consolidated consumer base that had been acquiring informal LPG before their incursion into this activity. With their increasing diversification into LPG, criminal groups are finding another black-market enterprise exploiting Mexico's most vulnerable, profiting from a product that has an inelastic demand, and that allows them to further consolidate their networks in excluded areas. What this case illustrates is that diversification is one of the main sources of resilience for criminal networks, allowing them to maintain their security networks, local area control and to finance their operating structures (Fuerte Celis et al. 2018, 9). Moreover, this is yet another instance of a long running trend in Mexican criminality: to guarantee resources and continuity in a hostile environment, criminal networks have continuously developed new specialized resource-extracting structures (ibid., 16).

The MFBM After One Year of the Crackdown

Having the opportunity to analyse the behaviour of the MFBM after a one-year crackdown leaves us with important insights. First, the black market's recent history points that it has experienced state repression and that, despite these interventions, this enterprise found ways to continue and expand. Second, the *obradorista* crackdown has decreased illegal extraction points in pipelines for the first time since 2004. Yet the local scenarios behind this decrease point to a mixed situation rather than the overwhelming victory: fuel-theft hotspots have reconfigured, and some have shown concerning expansions (Hidalgo and Estado de México), while in others the reduction call for cautious optimism (Puebla and Tamaulipas).

Then there is Guanajuato, a state where a considerable drop in illegal extraction points has taken place in parallel to a significant increase in homicidal criminal violence and the expansion of predatory illicit activities. These complex situations all combine with longstanding security limitations of the state to make a cautionary point about MFBM reductions: conjunctural national-level decreases should not lead to anticipated triumphalism as it once did with the homicide decreases at the beginning of the EPN administration (where homicides were declining in certain hotspots but increasing in others) and that local dynamics must be considered.

The President's discourse seems more concerned with awarding its government a much-needed victory regarding the top priority of his administration: the rescuing of PEMEX embattled by multiple deep-rooted crises. The fight against fuel-theft was originally presented by this presidency as one of the five measures that would help improve PEMEX's critical financial situation. However, the results presented by the government had no impact on the SOE's finances. Adding to this, while the current presidency promoted an overwhelming and easy victory over the MFBM, PEMEX in other instances recognized the magnitude of the challenge it faces. In late 2018, PEMEX in its annual report to the US Securities and Exchange Commission acknowledged the challenge posed by the "*sophistication and scope*" ("Annual Report Pursuant..." 2018, 161) of the illegal networks involved in fuel trafficking.

This case against anticipated triumphalism gathers more force once we consider the problematic data that the Mexican government uses to claim its victory over the trafficking of oil derivatives; basing its accomplishments on only one fuel-theft source and the contradictory losses of the Custody Transfer metering system that more than triple those presented by the presidency. Later, an argument is made that the government is not undertaking measures to tackle the state and non-state actors (particularly those within PEMEX) that enabled the MFBM to thrive. All these omissions lead to the point that the state lacks a monitoring and measuring capacity over its energy installations and that the manifestation of these limitations is embodied in the closure of pipelines as a costly strategy of fuel-theft reduction.

The pipeline closure points to a more concerning trend of the Obrador presidency. The government is prioritizing temporary and costly measures over gradual and laborious institutional changes (Felbab-Brown 2019a, 11; Ríos 2020). If fuel-theft reductions are being achieved through pipeline closures, while neglecting the CItR of PEMEX and the participation of other non-state actors, we can argue (without even considering the government's data inconsistencies on the issue) that these decreases will not last. The lack of impactful measures, perpetuating long-standing institutional shortcomings, enhances the resilience of criminal networks that have shown their capacity to adjust to state intervention employing means such as mobility, co-option, redundancy and diversification.

A consequence of this diversification phenomenon following the obradorista crackdown has been the increased participation of criminal networks in LPG trafficking. LPG, besides being profitable, offers criminals several advantages incentivizing their participation in this black market: a larger pipeline network with more legal players involved, lower prosecution rates and more impunity than other MFBM operations. The LPG black market, like its other MFBM counterparts, operates utilizing illegal camouflaging and co-option of grey actors that allow criminal networks to occupy spaces in institutional and economical legal realms increasing their resilience. LPG also allows criminal groups to consolidate their presence in lower-income areas where state presence is weaker, illegal enterprise diversification is easier, and there is a presence of robust informal economies to launder profits. Adding to these benefits, the LPG sector in Mexico has historically had a robust informal branch facilitating criminal co-option.

All these reasons make a case against anticipated official triumphalism. Structural shortcomings linked to Mexico's intensifying security crisis are deep-rooted, and they will surely not go away with celebratory discourses. The MFBM is manifestation of a country that has become more tragically criminal and violent for more than a decade. I won't disappear anytime soon.

Bibliography

"Alejandro Madrazo: La Guardia Nacional no dará federalism." 2019. Video. https://www.youtube.com/watch?v=yRO1LrR7yB8. Date accessed: July 9, 2021.

"Continúa reclutamiento nacional para policía municipal de Irapuato." *Bajío Web*, July 31, 2019. https://bajioweb.com/continua-reclutamiento-nacional-para-policia-municipal-de-irapuato/. Date accessed: July 9, 2021.

"El nuevo reto para el Gobierno de AMLO es el robo de gas; CJNG y Santa Rosa disputan el "negocio."" *Sin Embargo*, October 30, 2019. https://www.sinembargo.mx/30-10-2019/3670414. Date accessed: July 9, 2021.

"En Pemex se conjugan prácticas de corrupción y delincuencia organizada que ameritan investigaciones: Ana Lilia Pérez." *Aristegui Noticias*, February 28, 2020. https://aristeguinoticias.com/2802/mexico/en-pemex-se-conjugan-practicas-de-corrupcion-y-delincuencia-organizada-que-ameritan-investigaciones-ana-lilia-perez/. Date accessed: July 9, 2021.

"FORTASEG 2020: el Monto Más Bajo en los Últimos 8 Años." *Consensa*, October 19, 2019.

"Fuerzas de Seguridad Pública del Estado." Secretaría de Seguridad Pública del Estado de Guanajuato (SSPE). Accessed on April 30, 2020. http://seguridad.guanajuato.gob.mx/fspe/. Date accessed: July 9, 2021.

"Impera en 4 Estados 'ordeña' de ductos, confirma Pemex." *Vanguardia*, November 17, 2015. https://vanguardia.com.mx/articulo/impera-en-4-estados-ordena-de-ductos-confirma-pemex. Date accessed: July 9, 2021.

"La guerra que sigue: 2019 el año más violento en México." *Animal Político*, April 23, 2020. Video. https://www.animalpolitico.com/video/la-guerra-que-sigue-2019-el-ano-mas-violento-en-mexico/. Date accessed: July 9, 2021.

"Los Sinaloa y CJNG se pelean la plaza de gas LP en Puebla capital." *Diario Cambio*, March 12, 2019. https://www.diariocambio.com.mx/2019/secciones/codigo-rojo/item/8083-los-sinaloa-y-cjng-se-pelean-la-plaza-de-gas-lp-en-puebla-capital. Date accessed: July 9, 2021.

"Pemex investiga a sus empleados." *Expansión*, July 30, 2009. https://expansion.mx/actualidad/2009/07/30/pemex-investiga-a-sus-empleados. Date accessed: July 9, 2021.

"Pérdidas de Pemex casi se duplicaron en 2019." *Milenio*, February 27, 2020. https://www.milenio.com/negocios/pemex-perdidas-aumentaron-92-2019-ventas-cayeron. Date accessed: July 9, 2021.

"Policía Federal despliega operativo de seguridad en poliducto de Pemex." *Gobierno Federal*. Boletin 361. July 15, 2015.

"Puebla, una de las seis entidades de la República en alerta por el robo de gas LP." *Página Negra*, November 18, 2019. https://www.periodicocentral.mx/2019/pagina-negra/huachicol/item/26280-puebla-una-de-las-seis-ent

idades-de-la-republica-en-alerta-por-el-robo-de-gas-lp. Date accessed: July 9, 2021.
"Reapertura de poliducto aumentará violencia en Texmelucan." *Municipios Puebla*, January 10, 2020. https://municipiospuebla.mx/nota/2020-01-10/san-mart%C3%ADn-texmelucan/reapertura-de-poliducto-incrementar%C3%A1-violencia-en-texmelucan. Date accessed: July 9, 2021.
"Rechazan gasolineros verificaciones de PFP." *El Siglo de Torreón*, August 13, 2004. https://www.elsiglodetorreon.com.mx/noticia/103818.rechazan-gasolineros-verificaciones-de-p.html. Date accessed: July 9, 2021.
"Reditúa a Pemex acción antirrobo." *Reforma*, March 28, 2004.
"Varados, 24 buques en Veracruz; suman hasta un mes sin descargar combustibles." *Excelsior*, January 11, 2019. https://www.excelsior.com.mx/nacional/varados-24-buques-en-veracruz-suman-hasta-un-mes-sin-descargar-combustibles/1289625.. Date accessed: July 9, 2021.
Alire, David. 2019. "Mexico's Pemex eyes new oilfields after another year of big losses." *Reuters*, February 27. https://www.reuters.com/article/us-pemex-results/mexicos-pemex-posts-7-6-billion-loss-in-2018-as-oil-output-dips-idUSKCN1QG2G5.. Date accessed: July 9, 2021.
Ángel, Arturo. 2019. "Ocho masacres y aumento de la violencia en 19 estados durante primer año del gobierno de AMLO." *Animal Político*, December 4. https://www.animalpolitico.com/2019/12/masacres-aumento-violencia-gobierno-amlo/.. Date accessed: July 9, 2021.
Auditoría Superior de la Federación. 2016. "Pemex Transformación Industrial Producción de Gasolinas Auditoría de Desempeño: 16–6–90T9M-07–0482 482-DE," 1–25. https://www.asf.gob.mx/Trans/Informes/IR2016b/Documentos/Auditorias/2016_0482_a.pdf. Date accessed: July 9, 2021.
Ayling, Julie. 2009. "Criminal Organizations and Resilience." *International Journal of Law, Crime and Justice* 37: 182–196.
Beauregard, Luis Pablo. 2020. "2019 se convierte en el año más violento en la historia reciente de México." *El País*, January 20. https://elpais.com/internacional/2020/01/21/mexico/1579621707_576405.html. Date accessed: July 9, 2021.
Calderón, Gabriela, Beatriz Magaloni, Alberto Diaz-Cayeros, Gustavo Robles, and Jorge Olarte. 2015. "The Beheading of Criminal Organizations and the Dynamics of Violence in Mexicos Drug War." *Journal of Conflict Resolution* 59 (8): 1455–1485.
Carretto, Bianca. 2019. "El gobierno de López Obrador prevé reabrir siete ductos de hidrocarburos." *Expansión*, January 14. https://politica.expansion.mx/presidencia/2019/01/14/el-gobierno-de-lopez-obrador-preve-reabrir-siete-ductos. Date accessed: July 9, 2021.
Cacho Carranza, Yureli (2019). "Robo de gas LP Gran problemática nacional", Petroquimex, 4 de diciembre de 2018, pp. 38–43.

Centro de Investigación para el Desarrollo, A.C. (CIDAC). 2015. Es Posible Desarrollar Un Mercado De Gas LP Competitivo En México," 5–63. Mexico City: CIDAC.

Cruz Serrano, Noé. 2020. "PEMEX, víctima de huachicoleo al por mayor." *El Universal*. February 19, 2020. https://www.eluniversal.com.mx/cartera/sufrio-pemex-huachicoleo-al-por-mayor-durante-2019. Date accessed: July 9, 2021.

De la Rosa, Tomás. 2019 "Desabasto de gasolina con impacto de 0.49% al pib." *Eje Central*, February 15. https://www.ejecentral.com.mx/desabasto-gasolina-impacto-0-49-del-pib/. Date accessed: July 9, 2021.

Durán-Martínez, Angélica. 2018. *The Politics of Drug Violence: Criminals, Cops and Politicians in Colombia and Mexico*, 1–299. Oxford University Press.

EnergeA and Grupo Atalaya. 2017. "Estudio para analizar la problemática de seguridad física en las instalaciones del sector hidrocarburos y emitir recomendaciones para el reconocimiento de costos por concepto de seguridad que la Comisión Reguladora de Energía lleva a cabo en sus procesos de revisión tarifaria", pp. 1–368.

Farfán, Méndez Cecilia. 2019. "Beyond the War on Drugs: Violence and Security in Mexico." In Human Security and Chronic Violence in Mexico: New Perspectives and Proposals from Below, edited by Gema Kloppe-Santamaría and Alexandra Abello Colak, 57–81. Miguel Ángel Porrúa.

Felbab-Brown, Vanda. 2019a. "Mexico's Out-Of-Control Criminal Market," 1–29. The Brookings Institution.

Felbab-Brown, Vanda. 2019b. "AMLO's Security Policy: Creative Ideas, Tough Reality," 1–38. The Brookings Institution.

Flores, Nancy. 2019. "Inservibles 170 sitios estratégicos de monitoreo anti-huachicol en PEMEX." *Contralínea*, February 18. https://contralinea.com.mx/inservibles-170-sitios-estrategicos-de-monitoreo-antihuachicol-en-pemex/. Date accessed: July 9, 2021.

Fuerte Celis, María del Pilar, Enrique Pérez Lujan, and Rodrigo Cordova Ponce. 2018. "Organized crime, Violence, and Territorial Dispute in Mexico (2007–2011)." *Trends in Organized Crime* 22: 188–209.

Garay-Salamanca, Luis, Eduardo Salcedo-Albarán, Guillermo Macías Fernández, Diana Santos Cubides, and Nathalia Guerra Villamizar. 2018. "Macro-Corruption and Institutional Co-optation: The "Lava Jato" Criminal Network." Vortex Foundation, 13–209.

García, Karol. 2019. "Robo de gas LP restaría a Pemex 5,580 mdp este 2019." *El Economista*, October 23. https://www.eleconomista.com.mx/empresas/Robo-de-gas-LP-restaria-a-Pemex-5580-mdp-este-2019-20191023-0056.html. Date accessed: July 9, 2021.

García, Karol. 2020. "Pemex dejó de vender más que lo "recuperado" por robo." *El Economista*. January 28. https://www.eleconomista.com.mx/emp

resas/Pemex-dejo-de-vender-mas-que-lo-recuperado-por-robo-20200128-0013.html. Date accessed: July 9, 2021.

Garduño, Roberto, and Enrique Méndez. 2007. "Destino sin aclarar de 31 millones 387 mil barriles de petróleo en 2005: ASF." *La Jornada*, May 16. https://www.jornada.com.mx/2007/05/16/index.php?section=economia&article=023n1eco. Date accessed: July 9, 2021.

Gobierno Fácil Puebla. 2020. "Robo de Combustible en Puebla, del año 2000 a 2019." http://gobiernofacil.com/proyectos/robo-de-combustible-en-puebla. Date accessed: July 9, 2021.

González, Nayeli. 2019. "Sale más caro transportar el combustible en pipas que por ductos." *Excelsior*, January 7. https://www.dineroenimagen.com/economia/sale-mas-caro-transportar-el-combustible-en-pipas-que-por-ductos/105988. Date accessed: July 9, 2021.

Guerrero Eduardo. 2018a. "La segunda ola de violencia." *Nexos*, April 1. https://www.nexos.com.mx/?p=36947. Date accessed: July 9, 2021.

Guerrero Eduardo. 2018b. "Seguridad, ¿hasta cuándo?." Nexos. January 1, 2018. https://www.nexos.com.mx/?p=35375&fbclid=IwAR3aZTKJ87ci2GkCifcXeKxSrSwPOJW7j0zdxO-XmSo40BvSvY4jlpbP2M. Date accessed: July 14, 2021.

Guerrero Eduardo. 2020. "Cambio de rumbo, error de cálculo." *Nexos*, February 1. https://www.nexos.com.mx/?p=46652. Date accessed: July 9, 2021.

Hernández, Ibarzábal José Alberto, and David Bonilla. 2020. "Examining Mexico's energy policy under the 4T," 1–7. The Extractive Industries and Society, 17 March.

Hernández, Marlen. 2020. "Buscará Pemex reabrir 5 ductos." Reforma. January 8. https://www.reforma.com/buscara-pemex-reabrir-5-ductos/ar1848203?referer=--7d616165662f3a3a6262623b727a7a7279703b767a783a--. Date accessed: July 9, 2021.

Hope, Alejandro. 2019. "Minatitlán y la lógica de las masacres." *El Universal*, April 22. https://www.eluniversal.com.mx/columna/alejandro-hope/nacion/minatitlan-y-la-logica-de-las-masacres. Date accessed: July 9, 2021.

Hope, Alejandro. 2020a. "La Guardia Nacional y la Danza de los Números." *El Universal*, February 17. https://www.eluniversal.com.mx/opinion/alejandro-hope/la-guardia-nacional-y-la-danza-de-los-numeros. Date accessed: July 9, 2021.

Hope, Alejandro. 2020b. "No, los homicidios no están disminuyendo." *El Universal*, March 3. https://www.eluniversal.com.mx/opinion/alejandro-hope/no-los-homicidios-no-estan-disminuyendo. Date accessed: July 9, 2021.

Human Rights Watch. 2017. "World Report 2017: Mexico." https://www.hrw.org/world-report/2017/country-chapters/mexico. Date accessed: July 9, 2021.

Instituto Nacional de Geografía y Estadística (INEGI). 2019. "Encuesta Nacional de Victimización y Percepción sobre Seguridad Pública 2019." Instituto Nacional de Estadística y Geografía (INEGI). September 24. https://www.inegi.org.mx/contenidos/programas/envipe/2019/doc/envipe2019_presentacion_nacional.pdf. Date accessed: July 9, 2021.

Instituto para la Gestión, Administración y Vinculación Municipal (IGAVIM). 2020. "Robo de Hidrocarburo y Gas LP en Ductos y Pipas que Transportaban Hidrocarburo y Gas LP." IGAVIM.

Jiménez, Benito. 2018. "Indagan por ordeña a mandos en Pemex." *Reforma*, May 25. https://www.reforma.com/aplicacioneslibre/articulo/default.aspx?id=1403058&md5=4f823a4ee38d65d0ac1c18ec523d259c&ta=0dfdbac11765226904c16cb9ad1b2efe&lcmd5=7f706e0dbb5d692c0fc8c6b7bd20b171. Date accessed: July 9, 2021.

Jiménez, Benito. 2019a. "Se dispara en 318% robo de gaschicol." *Reforma*, November 23.

Jiménez, Benito. 2019b. "Padece Texcoco a cártel de gas LP." *Reforma*, August 12.

Kleiman, Mark. 2011. "Surgical Strikes in the Drug Wars." *Foreign Affairs*, September. October 2011. https://www.foreignaffairs.com/articles/north-america/2011-09-01/surgical-strikes-drug-wars. Date accessed: July 9, 2021.

León, Sáez Samuel. 2019a. "El "otro dato" del huachicol (parte I)." *Nexos*, September 30. https://datos.nexos.com.mx/?p=1028. Date accessed: July 9, 2021.

León, Sáez Samuel. 2019b. "El "otro dato" del huachicol (parte II)." *Nexos*, October 16. https://datos.nexos.com.mx/?p=1044. Date accessed: July 9, 2021.

Lessing, Benjamin. 2015. "Logics of Violence in Criminal War." *Journal of Conflict Resolution* 59 (8): 1486–1516.

Leyva, Trujillo Marlon, and Arguello Sergio Israel Salazar. 2017. "Emisiones, fugas y derrames en el transporte de hidrocarburos por ductos." Undergraduate Thesis, National Autonomous University of Mexico, 10–203.

Llano, Manuel, and C. Flores. 2017. "Ductos, ¿por dónde circulan los hidrocarburos en México?" [map]. Scale 1:3,500,000. México: CartoCrítica / Fundación Heinrich Böll.

López, Alejandra. 2012. "Resalta gas LP por clandestino." *El Norte*, December 11. https://www.negocioselnorte.com/aplicaciones/articulo/default.aspx?id=99887&v=2. Date accessed: July 9, 2021.

López, Ana, Maximilian Holst, and Magda Ramírez. 2020. "Infographic: Guanajuato: Territorio en Disputa." *México Evalúa*, February 4.

López, Jorge X. 2019. "Se roban 75 MDD, a la semana, en crudo." *24 horas*, January 24. https://www.24-horas.mx/2019/01/24/se-roban-75-mdd-a-la-semana-en-crudo-infografia/. Date accessed: July 9, 2021.

Madrid Ayala, Víctor Manuel, Maricela Gómez, Carlos Aguilar, José Antonio Márquez, and Denis Figueroa. 2018. "Transición Hacia Mercados Competidos De Energía: Gas LP." Cuadernos De Promoción De La Competencia, 9–87. Mexico City: Comisión Federal de Competencia Económica.

Medellín, Jorge Alejandro. 2005. "Pemex: la vaca negra... 'ordeña' clandestina y robo." *El Universal*. March 21, 2005.

Meyer, Lorenzo. 2019. "PEMEX o no PEMEX, esa (no) es la cuestión." *El Universal*, May 12. https://www.eluniversal.com.mx/columna/lorenzo-meyer/nacion/pemex-o-no-pemex-esa-no-es-la-cuestion. Date accessed: July 9, 2021.

Meza, Orozco Nayeli. 2020. "Huachicol: Saldos e Incógnitas del Combate." *Reporte Indigo*, January 16. https://www.reporteindigo.com/reporte/huachicol-saldos-e-incognitas-del-combate/. Date accessed: July 9, 2021.

Montero Vieira, José Ignacio. 2016, June. "El robo de combustible en México en el contexto del narcotráfico: Una vía alternativa de financiación," 1–15. Instituto Español de Estudios Estratégicos.

Morales, Isidro. 2020. "The Future of Pemex: Return to the Rentier-State Model or Strengthen Energy Resiliency in Mexico?" 1–45. Rice University's Baker Institute for Public Policy.

Moreno, Ana Lilia. 2019. "El combate al robo de combustible y la (incompleta) rendición de cuentas." *Animal Político*. https://www.animalpolitico.com/lo-que-mexico-evalua/el-combate-al-robo-de-combustible-y-la-incompleta-rendicion-de-cuentas/. Date accessed: July 9, 2021.

Nájar, Alberto. 2019. "Pemex: "Gaschicol", el nuevo y productivo negocio de los carteles en México." *BBC*, October 10. https://www.bbc.com/mundo/noticias-america-latina-49942340. Date accessed: July 9, 2021.

Nava, Diana. 2019. "El 8% del gas LP que se comercializa en el país es robado: Amexgas." *El Financiero*, October 23. https://www.elfinanciero.com.mx/economia/el-8-del-gas-lp-que-se-comercializa-en-el-pais-es-robado-amexgas. Date accessed: July 9, 2021.

Nava, Diana. 2020. "¿Por qué Pemex perdió el grado de inversión y qué sigue?" *El Financiero*, April 18. https://www.elfinanciero.com.mx/economia/por-que-pemex-perdio-el-grado-de-inversion-y-que-sigue/. Date accessed: July 9, 2021.

Noticieros, Televisa. 2019. "¿Cómo explicar la pérdida millonaria de Pemex en 2019? - Es la hora de opinar." Video. https://www.youtube.com/watch?v=J5xJf5PITBo&list=LLxV5tnKH6eIqSjnHM7YLv2g&index=17&t=0s. Date accessed: July 9, 2021.

Padilla Oñate, Sergio. 2019. "Los límites a la militarización de la seguridad pública en México." *Animal Político*, October 23. https://www.animalpolitico.com/el-blog-de-causa-en-comun/los-limites-a-la-militarizacion-de-la-seguridad-publica-en-mexico/. Date accessed: July 9, 2021.

PEMEX. "Reporte de tomas clandestinas en 2018." https://www.pemex.com/acerca/informes_publicaciones/Paginas/tomas-clandestinas.aspx. Date accessed: July 9, 2021.
Pérez, Ana Lilia. 2011. "El Cártel Negro: Cómo el crimen organizado se ha apoderado de Pemex," 5–221. México: Grijalbo.
Pérez, Ana Lilia. 2012. "El sexenio de la ordeña." *Contralínea*. January 4. https://contralinea.com.mx/el-sexenio-de-la-ordena/. Date accessed: July 9, 2021.
Petróleos Mexicanos. 2018. "Annual Report Pursuant to Section 13 Or 15(D) Of the Securities Exchange Act of 1934," 3–432. United States Securities and Exchange Commission.
Radden Keefe, Patrick. 2013. "The Geography of Badness: Mapping the Hubs of the Illicit Global Economy." In *Convergence: Illicit Networks and National Security in the Age of Globalization*, edited by Michael Miklaucic and Jacqueline Brewer, 97–107. National Defense University Press.
Raziel, Zedryk. 2018. "Alista AMLO plan contra huachicol," *Reforma*. December 7. https://www.reforma.com/aplicaciones/articulo/default.aspx?id=1558407&v=7. Date accessed: July 9, 2021.
Rea, Daniela, and Pablo Ferri. 2019. "La Tropa: Por qué mata un soldado," 13–297. Penguin Random House.
Reforma. 2019. "Opera red de huachicoleo de NL a CDMX." Video. https://www.youtube.com/watch?v=UWQ00_XExQo. Date accessed: July 9, 2021.
Reza, Gloria. 2019. "CJNG se deslinda del huachicoleo y expresa apoyo a medidas del gobierno," *Proceso*, February 13. https://www.proceso.com.mx/nacional/2019/2/13/cjng-se-deslinda-del-huachicoleo-expresa-apoyo-medidas-del-gobierno-220211.html. Date accessed: July 9, 2021.
Ríos, Viridiana. 2020. "México reduce su corrupción, por las razones equivocadas." *El país*, March 11. https://elpais.com/elpais/2020/03/11/opinion/1583947118_098114.html. Date accessed: July 9, 2021.
Rodríguez, García Arturo. 2019. "Por huachicol detienen al general Sócrates Alfredo Herrera Pegueros, exjefe de seguridad de Pemex," *Proceso*. July 18. https://www.proceso.com.mx/nacional/2019/7/18/por-huachicol-detienen-al-general-socrates-alfredo-herrera-pegueros-exjefe-de-seguridad-de-pemex-228152.html. Date accessed: July 9, 2021.
Romero, Octavio. 2019. "Comparecencia del Ingeniero Octavio Romero Oropeza, Director General de Petróleos Mexicanos (PEMEX), ante las Comisiones Unidas de Energía e Infraestructura de la Cámara de Diputados LXIV Legislatura." Presentation. Cámara de Diputados, October 28.
Romero, Octavio. 2020. "Presidential Press Conferenece." Presentation. Palacio Nacional. January 7.
Rosas, Obed. 2019. "AMLO presume menos huachicol, pero dice que aún hay comunidades que lo apoyan." *Expansión*. May 8. https://politica.expansion.

mx/presidencia/2019/05/08/amlo-presume-menos-huachicol-pero-dice-que-aun-hay-comunidades-que-lo-apoyan. Date accessed: July 9, 2021.
Saucedo, David. 2019. "Guerra de cárteles hunde al golpe de timón y baña de sangre a Guanajuato." *Poplab*, September 15. https://poplab.mx/article/GuerradecarteleshundealgolpedetimonybaadesangreaGuanajuato. Date accessed: July 9, 2021.
Secretariado Ejecutivo del Sistema Nacional de Seguridad Pública (SESNSP). "Datos Abiertos de Incidencia Delictiva." *Open Data Base*. https://www.gob.mx/sesnsp/acciones-y-programas/datos-abiertos-de-incidencia-delictiva?state=published. Date accessed: July 9, 2021.
Sierra, Miguel. 2019. "Qué es y quién conforma el CDN, el poderoso Cártel del Noreste." *El Universal*, December 3. https://www.eluniversal.com.mx/estados/que-es-y-quien-conforma-el-cdn-el-poderoso-cartel-del-noreste. Date accessed: July 9, 2021.
Silva, Carlos, Catalina Pérez Correa, and Rodrigo Gutiérrez. 2017. "Índice de letalidad 2008-2014: menos enfrentamientos, misma letalidad, más opacidad." *Perfiles Latinoamericanos* 25 (50): 331–359.
Silva, Martha. 2020. "Contra aplastante realidad de la violencia en Guanajuato, solo declaraciones políticas, no acciones." *Poplab*, January 9.
Storr, Samuel. 2019. "Seguridad Pública Enfocada en el Uso de la Fuerza E Intervención Militar: La Evidencia en México 2006-2018," 100–105. Mexico City: Universidad Iberoamericana.
Transparency Response "1857000006220." PEMEX. February 25, 2020.
Transparency Response "1857000059819." PEMEX. October 29, 2019.
Transparency Response "1857200007220." PEMEX. February 5, 2020.
Transparency Response "1857200094319." PEMEX. February 5, 2020.
Valdivia, García Jorge. 2019. "Desabasto, oportunidades perdidas, mentiras y muerte en la guerra contra el huachicol." Universidad Jesuita de Guadalajara, October 30. https://analisisplural.iteso.mx/2019/10/30/desabasto-oportunidades-perdidas-mentiras-y-muerte-en-la-guerra-contra-el-huachicol/#_ftn24. Date accessed: July 9, 2021.
Velázquez, Edmundo. 2019. "Huachicoleros tienen paso libre en Juan Galindo, usan cerro de Necaxaltepetl para la extracción de hidrocarburo." *Página Negra*, May 14. https://www.periodicocentral.mx/2019/pagina-negra/huachicol/item/11060-huachicoleros-tienen-paso-libre-en-juan-galindo-usan-cerro-de-necaxaltepetl-para-la-extraccion-de-hidrocarburo. Date accessed: July 9, 2021.
Vicenteño, David. 2004a. "Toma la PFP más plantas de Pemex." *Reforma*, March 6.
Vicenteño, David. 2004b. "Alista PFP nueva fase de operativo en Pemex." *Reforma*, March 28.

Zenteno, Jessica. 2020. "En 2019, Puebla fue el estado con más robo de pipas de gas LP e hidrocarburo a nivel nacional." *Página Negra*, March 13. https://www.periodicocentral.mx/2020/pagina-negra/huachicol/item/5339-en-2019-puebla-fue-el-estado-con-mas-robo-de-pipas-de-gas-lp-e-hidrocarburo-a-nivel-nacional. Date accessed: July 9, 2021.

CHAPTER 7

Conclusion

How did fuel trafficking become a prominent criminal market in Mexico and what factors explain its rapid growth between 2011 and 2018? The MFBM consolidated during the past two decades. Research shows that this black-market went through three stages: First, its beginnings in the late 1990s as an emerging illicit activity, not yet part of the enterprise portfolio of Mexican criminal networks. The second stage takes place between the mid-2000s–2009, with the Burgos operation and the Zetas' pioneering of macro-criminal network involvement in fuel trafficking. Finally, a third stage begins approximately in 2011 when multiple criminal actors begin diversifying towards fuel trafficking because of three factors: the fragmentation and diversification of criminal networks, the co-option of grey actors and increasing fuel prices.

As Mexico´s criminal landscape became more fragmented and competed we witnessed an increasing diversification towards local criminal enterprises that have been showing growth since the mid-2000s. Transnational drug trafficking became an increasingly inaccessible enterprise for a growing number of criminal actors. This combined with a wider tendency to exploit local illicit markets by criminal networks of all capacities as they offered the following incentives: accessing sustained revenues, consolidating social support, laundering criminal profits, infiltrating the legal economy, and co-opting authorities and other grey actors. In this context, criminal diversification became a sound pursuit as the Mexican illicit landscape became increasingly contested and violent.

This criminal diversification towards fuel trafficking was magnified by the Co-opted Institutional Reconfiguration (CItR) of PEMEX through a process of Systemic Macro-corruption. It is true that other non-state actors played important roles (entrepreneurs for distribution and money-laundering). Still, the importance of co-opting PEMEX is crucial to the consolidation of a MFBM, which became a parallel structure in the SOE´s production segments of Industrial Transformation, Logistics and Commercialization. This was achieved through the far-reaching and sustained agreements inherent to CItR and the large-scale co-option of grey actors across PEMEX´s chain of command, from on-site workers to union leaders and high-ranking officials.

The MFBM is thus a consequence of the CItR of PEMEX in which the SOE´s norms and procedures were manipulated for sustained corruption schemes. Along the way, the MFBM became a priority for Mexico´s criminal networks. The scale of money-laundering, façade companies, complex illegality camouflaging, diversity of involved actors, amounts of fuels stolen, varied theft methods and large profits attest to this. Its importance is also reflected by the various grey actors involved in the MFBM who use their resources and their presence between legality and illegality for fuel trafficking to expand. The CJNG, Zetas and many other multi-sized criminal networks are currently participating in this newly significant criminal market, profiting from hydrocarbons trafficking in different scales and capacities, while reconfiguring the geographies of criminal violence in Mexico.

The rise of the MFBM is one of the most important manifestations of the decade-long deterioration of the country's security situation. This crisis is rooted in deep structural trends, including the Mexican state´s security deficiencies, the criminal fragmentation cycle, illicit enterprise diversification and ferocious territorial expansion. The Mexican state has proven on countless occasions that it does not possess the policing or the justice institutions to curve black markets effectively. This history has taught us many lessons about Mexican public security strategy. Mexico's last three Presidents have all pursued a disjointed strategy of institution building that lacks continuity. The Federal Police under Calderón, the National Gendarmerie under EPN, and Obrador´s National Guard as well as their obscure implementation of justice and local police reforms all attest to this discontinuity and highlight the need for adopting a continuous and consistent state security policy and for committing to painfully slow institution building processes.

The importance of this continuity cannot be stressed enough, especially if Mexico wants to decrease the lethal violence that has brought on the country's most violent period since the Mexican Revolution. The decrease of state lethality under AMLO and its paltry consequences on homicidal violence points to this pressing need more than ever: the state needs to engage in a long institutional overhaul of its security and justice apparatus if it intends to reduce the homicidal violence that has become ever more fragmented, contested, and brutal. Reducing the role of the state in contributing to lethal violence is important. Yet, its minimal impact points to the necessity of a commitment to long-term reforms and implementation processes. These reforms and institution building processes are important to curb not only homicidal violence, but also impunity and corruption.

Another implication of this research is that, by exploring the appearance and consolidation of the MFBM, we can understand the dynamics, trends and actors that have brought Mexico to its current public security situation. The diversified and highly competed criminal underworld points to the necessity of turning our attention to other black markets to explain how this public security crisis has gone beyond a simple war on narcotraffickers. Research has been done on this regard, but the drug war discourse still holds a lot of cultural and analytic attention, and this investigation aspires to contribute to a growing literature about highly diversified criminality. This research advocates for not ignoring local dynamics and their repercussions. Disregarding them can lead to incorrect diagnoses, poor policies, and tragic consequences. Security and policy assessments have been plagued by a centralized vision and, while tending to national-level trends is important, it cannot justify ignoring the crucial role that local dynamics play in Latin American public security matters.

This research is one of the first attempts to academically analyse fuel trafficking. It is true that this is an in-depth study focused on Mexico, yet its implications and lessons can help us elucidate similar cases in other countries and regions. Across Latin America we have witnessed a dispersion of illegal markets towards the extraction of natural resources. By developing a detailed study on this illicit enterprise, this investigation offers theoretical points on territorially embedded subnational black markets that exploit weak state structures through co-option, a problem that has become a common trait of the Latin American criminal underworld and a key factor behind the region's high levels of homicidal violence.

Criminal homicidal violence in Brazil, Venezuela Colombia, Mexico, and the Central America subregion is mostly driven by groups that pursue profits from illicit economies largely shaped by subnational territorial hegemony. Competition between these organizations drives *"the very high violence patterns in the region"* (Yashar 2018, 65) particularly in the enclaves where law and order institutions are ineffective and corrupt. This violence has become widespread and endemic, affecting the lives of many of Latin Americans. Given its pressing importance, further research is needed to elucidate this phenomenon.

This research on the MFBM helps by contributing to this body of research, exploring how its appearance and growth as a black-market affected the dynamics of criminal violence within Mexico while offering analytical tools for similar regional contexts. These criminal dynamics are marked by geography and black-market diversification, fragmentation and enhanced competition where state and non-state actors simultaneously take part. Groups that deploy violence often have a presence in the licit and the illicit realms. This duality has allowed them to have an influence on governance, the formal and the informal economy, the delivery of social services and the daily life of many of Latin Americans.

By developing this case study, we can explore in further detail a trend of diversification of criminal enterprises into extractive sectors in countries like Colombia (González Garzón 2015), Brazil, Indonesia, the Democratic Republic of Congo (Boekhout van Solinge 2014). In these black markets (timber, rare minerals) many of the characteristics developed in this research are also present such as legality and criminality interfaces, the camouflaging of illegality and the relation between violence and illicit enterprises (González Garzón 2015; Boekhout van Solinge 2014; Boekhout van Solinge et al. 2016). This investigation adds to this body of research by contributing a comprehensive analysis of how these dynamics can influence a wide range of extractive black markets.

As the first chapter of this research elaborates, illegal smuggling of oil and its derivatives is present across the world. Despite the particularities of each region and nation, this research presents a framework useful to analyse such cases. This framework develops the following guidelines for analysing fuel black markets:

1. Measure the size of a fuels black-market in volumetric losses (barrels and/or litres). Researchers should consider the accuracy of their sources and reinforce their findings through crosschecking. Define whether a fuel trafficking market has a transnational, national and/or local scope; whether it is operating in an energy producing, consuming and/or transporting hub; and if subsidies, tariffs and/or taxes play a role in generating incentives for an internal or transborder black market. Identify whether illegal hydrocarbons markets rely on (1) stolen products (oil, derivatives) and/or (2) adulterated formulas or artisanal refined products.
2. Define which criminal networks are involved in fuel-theft and trafficking (operational capacity, territorial scope, diversification into other black markets, sophistication of their fuel-theft and trafficking operations). Identify fuel-theft methods: stealing tanker trucks, marine piracy, maritime bunkering, refinery-theft, theft in pipelines, distribution terminals (marine and inland), marine platforms and/or transborder trafficking. Determine whether there is a predominant source of stolen derivatives or if there are multiple sources. Clarify whether a fuel black-market is dependent on infrastructure (energy installations, mobility infrastructure, urban centres) or takes part in inaccessible areas.
3. Identify the grey actors (state and non-state) connecting the realms of legality and illegality. Define the functions they fulfil in the hydrocarbons black-market (money laundering, transport, guaranteeing impunity, distribution camouflaging, energy installations access, documents, offering market expanding opportunities, operational scope).
4. Explain how illegality camouflaging operates to protect fuel trafficking operations (extraction, transport, distribution and commercialization). Regarding distribution and commercialization, determine whether it takes place at the retail, wholesale level or both and define the locations where it occurs (highway sales, camouflaged or legal points or retail).
5. Define the actors that make up the demand side of the fuel black-market (local populations and/or large-scale consumers).

These points can help as a starting point for other individuals interested in analysing other cases of fuel trafficking. They are not exhaustive, and further research should strive to make contributions to enrich them.

This research also contributes to recognizing the state not as a unitary entity, but as a site of struggle where agents and entire institutions can be criminally co-opted. This co-optation does not aspire to topple the state but to protect interests (guaranteeing profits, ensuring impunity, extracting resources). In most cases where non-state criminal actors invade state functions, they are pursuing such objectives. Academics must challenge discourses that promote a separation between legal and illegal. Illegality and lawfulness operate on many occasions in tandem and, while areas and instances of confrontation do exist, it is important to highlight this problematic relationship between the two. This is such a complex dynamic that it transcends the exclusive co-optation of state institutions: illegality can tap into legality through front companies or non-governmental organizations and occupy grey areas that are not purely criminal, like informality. This intertwined relationship is troublesome and intricate, and promoting notions of separateness obstructs objective analysis and hinders finding solutions.

Another contribution of this research has to do with the relationship between the thriving criminal networks and development. The MFBM depends on the presence of energy-heavy industrial sectors, large infrastructure projects creating energy demand, communication infrastructure and financial tools that facilitate money-laundering. The MFBM offers an example of how modern-day macro-criminality depends on the presence of relatively consolidated states, development, and access to different services (telecoms, banking). These requirements also apply in marginalized areas, impoverished communities, and large informal economies. Recognizing these criminal hub contexts and requirements will help build more coherent policies to tend to modern criminal phenomena.

As mentioned previously, this is one of the first academic treatments of a fuel black-market case. At a Mexican level, it is my hope that this effort can lead to further research which could develop other state-level case studies (like Veracruz and Tamaulipas, for example), analyse the supply chain that fed demand, develop the export side of the black market, explore the history of this criminal phenomenon more profoundly, delve into the technical limitations within PEMEX that supported the expansion of the MFBM, and so on. The operations of maritime fuel-theft in the Gulf of Mexico and their diversification into piracy and extraction of platform equipment are other issues that could lead to interesting research.

At the international level, some valuable research to be done is the academic analysis of a transnational fuel trafficking operation (a detailed treatment of the Libya–Malta–Italy operation explored in this research, for example). Another similar case would be the transborder trafficking taking place between Venezuela and Colombia. In-depth investigations of countries that present large-scale operations of fuel trafficking (Nigeria, Venezuela, Poland, Libya and Turkey) would also be essential. Researching these cases in depth would help set a base of research that would allow for enlightening comparative studies.

BIBLIOGRAPHY

Boekhout van Solinge, Tim. 2014. "The Illegal Exploitation of Natural Resources." In *The Oxford Handbook of Organized Crime*, edited by Letizia Paoli, 500–526. Oxford University Press.

Boekhout van Solinge, Tim, Zuidema Pieter, Vlam Mart, Cerutti Paolo Omar, and Yemelin Valentin. 2016. "Organized Forest Crime: A Criminological Analysis with Suggestions from Timber Forensics." International Union of Forest Research Organizations, pp. 81–96.

González Garzón, Hermann David. 2015. "Trasgresión de derechos humanos a raíz del tráfico ilegal de coltán en el Departamento del Guainía." *Revista Científica De La Escuela De Postgrados De La Fuerza Aérea Colombiana* 10: 151–168.

Yashar, Deborah J. 2018. *Homicidal Ecologies: Illicit Economies and Complicit States in Latin America*, 1–368. Cambridge University Press.

Index

B
barrels, 19, 61, 69, 72, 79, 81–84, 86, 88, 91, 120, 130, 146, 157, 177, 185, 186, 217
black markets, 3–5, 7–11, 16, 17, 23, 24, 26–30, 34, 36–38, 40, 41, 49, 64, 67, 78, 95, 109, 111, 112, 116, 145, 149, 154, 155, 160, 161, 171, 214–217
Burgos Basin, 33, 68, 70–72, 74–78, 113, 115, 122, 178

C
Calderón, Felipe, 25, 64, 65, 70, 111, 130, 171, 172, 174, 178, 214
camouflaging, 8, 11, 22, 23, 29, 30, 33, 36, 43, 46, 49, 69, 92, 94, 96, 122, 128, 199, 200, 203, 214, 216, 217
Cartel Jalisco Nueva Generación. See CJNG
CJNG, 115, 118–121, 143, 145–147, 149, 150, 155, 157, 161, 162, 171, 182, 195, 214

Colombia, 4, 11, 16, 18, 21, 22, 114, 173, 216, 219
condensate, 16, 70–72, 74–77
co-option, 4, 5, 8, 9, 30–33, 36, 39, 40, 50, 59, 68, 78, 84, 88, 109, 115, 118, 122, 147, 153, 158, 159, 161, 172, 194, 203, 213, 214, 216
corruption, 4, 31–33, 59–62, 65, 67, 89, 123, 124, 128, 142, 160, 187, 200, 214, 215
criminal networks, 5, 7–9, 12, 15, 18, 20, 23–27, 32, 33, 43, 46, 48, 49, 67, 68, 70, 71, 77–79, 83, 84, 90, 92, 94–99, 109–111, 113, 114, 116–120, 122, 127, 128, 131, 141, 144, 145, 149, 150, 155–157, 160–162, 169, 173, 176, 177, 193–196, 199–201, 203, 213, 214, 217, 218
Custody Transfer metering system. See SITRAC

D

demand, 30, 37, 62, 76, 79, 92, 127, 130, 131, 143, 155, 160, 170, 171, 184, 192, 201, 217, 218

Democratic Revolution Party. *See* PRD

derivatives, 16, 19, 22, 23, 25, 30, 35, 48, 69–72, 78–80, 89, 93, 99, 126, 141, 155, 170, 177, 196, 202, 216, 217

diesel, 17–19, 68, 70–72, 80, 93, 95, 126, 127, 130, 131, 155, 156, 170, 178, 185, 186, 195, 196

distribution terminals, 16, 26, 81, 83, 86, 90, 125, 176, 183, 217

diversification, 3, 5, 7–11, 23–27, 33, 38–40, 43, 49, 64, 67, 78, 90, 96, 97, 99, 109–117, 120–122, 182, 194–196, 198, 201, 203, 213, 214, 216–218

drug trafficking, 27, 39, 62, 64, 65, 70, 72, 97, 110, 112, 115, 116, 119, 121, 131, 173, 213

E

Enrique Peña Nieto. *See* EPN

entrepreneurs, 20, 21, 41, 50, 72, 99, 214

EPN, 6, 62, 66, 67, 111, 125, 130, 131, 141, 171, 172, 174, 176, 178, 202, 214

EU, 16, 17, 19, 21, 23, 30, 37, 45

European Union. *See* EU

extraction points, 5, 68, 70, 78, 79, 85, 115, 116, 125, 144, 150, 152, 156, 158, 178–182, 184, 185, 192, 195, 196, 198, 202

extractive, 3, 4, 33, 49, 113, 216

F

federal, 64, 68, 83, 88, 93, 110, 150, 153, 158, 160, 162, 172, 174–178, 182, 183, 185–187, 192, 194, 196, 200

fragmentation, 5, 8, 9, 24–27, 33, 34, 41, 49, 63, 64, 66, 67, 109–113, 118, 145–147, 174, 213, 214, 216

fuel smuggling, 7, 8, 16–19, 21, 30, 33, 38, 40, 42–44, 68, 72, 79, 154

fuel-theft, 7, 16, 19, 22, 24, 68–70, 72, 78, 80, 81, 84, 85, 87, 88, 94, 115, 116, 119, 120, 123–125, 128, 143–145, 150, 152, 155, 157–160, 173, 177, 178, 182–185, 189, 202, 203, 217, 218

fuel trafficking, 3–9, 11, 12, 15–24, 27, 29, 30, 33, 34, 36–44, 46, 48–50, 59, 67–72, 74–81, 83, 85, 90, 92, 94, 111, 113, 115–128, 131, 142, 143, 145, 149, 151, 152, 154–162, 171, 177, 178, 182, 184, 185, 187, 192–196, 202, 213–215, 217, 219

G

gasoil, 40, 41, 43–46

gasoline, 17, 18, 22, 23, 68, 80, 93, 116, 126, 130, 131, 170, 175, 178, 185, 186, 195, 196

GC, 64, 72, 75–77, 112, 116, 145, 147

grey actors, 9, 11, 20–23, 30, 31, 33, 37, 39, 40, 43, 46, 49, 75, 78, 85, 87, 88, 98, 99, 109, 122, 123, 126–128, 142, 149, 158–160, 173, 187, 199, 203, 213, 214, 217

grey agents. *See* grey actors
Guanajuato, 6, 9, 27, 63, 78, 120, 124, 141–147, 149, 150, 160–162, 172, 177, 178, 182, 191, 192, 195, 202
Gulf Cartel. *See* GC

H
highways, 68, 95, 113, 117, 118, 151
homicides, 65–67, 83, 119, 144, 145, 157, 171, 176, 182, 202

I
illicit networks, 3, 19, 36, 110, 115, 201
incentives, 3, 8, 17, 94–96, 111, 141, 213, 217
Institutional Revolutionary Party. *See* PRI
Ireland, 18, 77
Italy, 19, 39, 43, 46, 189, 219

L
La Familia Michoacana, 111, 115, 118, 145, 147
Latin America, 8, 10, 11, 15, 17, 21, 93, 95, 97, 110, 118, 200, 215
Libya, 8, 19, 21, 33–37, 39–43, 46, 49, 219
Liquified Petroleum Gas. *See* LPG
LPG, 170, 194–196, 198–201, 203

M
Magna, 128, 131, 185, 186
Malaysia, 17, 18
Malta, 8, 19, 21, 33, 37, 38, 41, 43, 44, 48, 219
marine, 16, 21, 38, 40, 80, 92, 93, 120, 175, 176, 191, 217

maritime, 21, 36, 37, 43, 45, 90–93, 217, 218
Mexican Fuel Black Market. *See* MFBM
Mexico, 3–11, 16, 23–27, 32, 33, 40, 49, 50, 59–69, 71, 72, 75, 78–80, 86, 87, 90–93, 95, 97, 109, 110, 112, 113, 115, 117–119, 123, 124, 126–128, 130, 141, 143, 145–147, 149–152, 154, 156, 158, 160–162, 169–177, 180, 182, 185, 192, 196, 199–201, 203, 213–216, 218
MFBM, 5, 8, 9, 11, 15, 25, 27, 31, 32, 34, 50, 59, 68, 70, 78–83, 92, 94–96, 98, 99, 109, 112, 116, 117, 119, 120, 122, 123, 125–128, 131, 141, 143–146, 150, 151, 153–158, 160–162, 169, 171, 176–178, 180–186, 189, 191–196, 198, 200–203, 213–216, 218
money-laundering, 3, 20, 23, 30, 31, 39, 48, 49, 72, 83, 93, 113, 121, 122, 127, 146, 159, 160, 173, 200, 214, 218
municipal, 63, 65, 149, 150, 157–159, 161, 172–174, 200

N
narcotics, 18, 23, 24, 26, 28, 30, 39, 97, 110, 113, 114, 116, 122, 162
Nigeria, 16, 18, 20–22, 30, 144, 219

O
Obrador, López, 4, 6, 9, 80–82, 85, 86, 123, 125, 126, 162, 169–173, 176–179, 184, 187, 189, 194, 203, 214

oil, 15, 16, 18, 19, 21, 30, 35, 37, 39, 41, 42, 59–62, 71, 77, 90–92, 111, 114, 123, 125, 130, 144, 169, 170, 177, 183, 185, 202, 216, 217

P

PAN, 63, 152, 159
Partido Acción Nacional. See PAN
PEMEX, 4, 6, 8, 31–33, 59–62, 68–72, 75–90, 92–96, 98, 99, 109, 116, 117, 122–128, 132, 147, 149, 169–171, 176–178, 182–187, 189, 191–193, 195, 196, 198, 202, 203, 214, 218
Petróleos Mexicanos. See PEMEX
pipelines, 5, 21, 25, 35, 68, 70, 79–81, 83–87, 89, 94, 98, 114, 116, 124, 125, 128, 142, 144, 149–151, 153, 156, 178, 180, 182, 183, 185, 186, 189, 191, 192, 195, 196, 198, 202, 203, 217
piracy, 16, 92, 113, 155, 157, 183, 217, 218
platforms, 90–92, 183, 217
Poland, 18, 219
police, 32, 63, 64, 67, 69, 149, 157–161, 171–174, 177, 200, 214
polyduct, 86, 92, 178, 189, 192, 198
PRD, 76, 150, 159
Premium, 80, 128, 131, 185, 186
PRI, 62, 63, 123, 152, 158, 159
prices, 9, 15, 17, 18, 60, 62, 93, 95, 128, 130, 131, 141, 156, 184, 186, 213
Puebla, 6, 9, 30, 115, 120, 124–126, 141, 142, 144, 145, 150–152, 154–162, 180, 191–193, 195, 196, 199, 202

R

refinery, 16, 19, 21, 26, 35, 36, 38, 40, 68, 80, 81, 83–85, 90, 92–94, 120, 122, 124, 125, 128, 142, 143, 145, 147, 149, 160, 176, 177, 180, 183, 217

S

Santa Rosa de Lima Carte. See SRLC
SCADA, 89, 90, 176, 189
security forces, 145, 149, 160, 174–176, 182, 194
ships, 21, 22, 34, 41, 43–45, 92, 93, 183
Sinaloa Cartel, 67, 117, 118, 120, 146, 155, 171, 195
SITRAC, 185, 186
SRLC, 143, 145–147, 149, 150, 161, 162, 195
state actors, 5, 8, 31, 32, 43, 48, 50, 83, 94, 99, 115, 127, 128, 202, 203, 214, 216
stations, 36, 37, 45, 70, 72, 79, 95, 96, 120, 125–127, 158, 160, 176, 177, 191
Supervisory Control and Data. See SCADA
supply, 3, 19, 22, 29, 30, 45, 46, 48, 79, 92, 96, 97, 110, 115, 127, 141, 155, 160, 191, 201, 218
Syria, 19, 20
Systemic Macro-corruption, 32, 122, 128, 189

T

Tamaulipas, 16, 64, 69, 72, 75, 76, 78, 92, 93, 96, 115, 116, 119, 124–126, 156, 177, 178, 180, 191, 195, 202, 218

tanker trucks, 21, 36, 37, 68, 71, 72,
 74–77, 81, 85, 88, 92, 93, 120,
 123, 156, 183, 192, 199, 217
Thailand, 17, 18, 21, 23
Turkey, 16, 17, 19, 21, 30, 219

V

Venezuela, 4, 18, 20, 22, 114, 216,
 219
violence, 10–12, 24–26, 34, 62–67,
 110, 112, 119, 141, 144–146,
 154, 157, 161, 162, 171, 172,
 174–176, 182, 192, 202,
 214–216

Z

Zetas, 27, 64, 72, 75–78, 92, 94–96,
 110–116, 118, 120, 121, 124,
 145–147, 150, 153, 155–157,
 162, 176, 195, 213, 214

The manufacturer's authorised representative in the EU is Springer Nature Customer Service Centre GmbH, Europaplatz 3, 69115 Heidelberg, Germany. If you have any concerns regarding our products, please contact ProductSafety@springernature.com

Printed and bound by CPI Group (UK) Ltd, Croydon, CR0 4YY

26/03/2026

02078942-0003